History of Mathematical Logic from Leibniz to Peano

History of Mathematical Logic from Leibniz to Peano

N. I. Styazhkin

The M.I.T. Press
Cambridge, Massachusetts, and London, England

Set in Monotype Baskerville.
Printed and bound in the United States of America by
The Colonial Press, Inc., Clinton, Massachusetts.

Originally published by Nauka (Moscow, 1964) as
Stanovleniye idey matematicheskoy logiki.

SBN 262 19057 5 (hardcover)

Library of Congress catalog card number: 69-14408

Preface

Two basic approaches have been used by researchers in studying the history of logic: the philological and the "retrospective–logical." The first approach entails careful terminological examination of source materials, comparison of different variants of the same documents, and clarification of the relationships between the logical facts and the data from the science of language. The second approach reduces to analysis of older logical concepts in order to find elements of contemporary viewpoints, using contemporary logical apparatus as a thread of Ariadne for guidance through the labyrinth of past logical investigations.

While both these approaches have undeniable advantages, they also have disadvantages. Historians who use the philological approach risk reducing their study to a history of terminological innovation, while researchers using the retrospective–logical method run the risk of modernizing the material under investigation. This book attempts to combine both approaches to avoid undesirable excesses. In considering older logical theories, the author has tried to explain how they have affected the corresponding modern viewpoints and simultaneously to devote the requisite attention to philological aspects.

This book deals with the development of a number of ideas and attitudes in mathematical logic, covering the period from the Middle Ages to the beginning of the twentieth century.

Throughout its entire history, mathematical logic has been closely related to development of other sciences, and its own development has been affected by the state of scientific methodology (primarily mathematics).

Mathematical examination of the deductive part of classical Aristotelian logic is the distinguishing characteristic in the works of the majority of the pioneers in the field of mathematical logic. The logical calculuses presented in these works are analogous to the well-known algebraic calculuses of Viète and Descartes. These early logical calculuses made use of two types of expressions, one called "subjects"; the other, "conclusions," "predicates," or "relations." When constant individual subjects were substituted for the corresponding variables in expressions of the second type, propositions characterizable by tests of truth or falseness were obtained. It was possible to use definite, rather simple rules reminiscent of the rules of arithmetic algebra to operate with propositions written in an artificial symbolic language.

Although the approach taken by Leibniz was not directly related to the requirements of the mathematics of his time, the proponents of this approach obtained individual valuable scientific results. His time also saw the actual ripening of an extremely fruitful notion in contemporary mathematics — the notion of a distributive structure, which must be dealt with in both logic and mathematics as well as in their applications to physics and other sciences. The appearance of this idea indicated a very deep penetration into the essence of logical and mathematical operations.

In conclusion, the author would like to take this opportunity to express his deep gratitude for helpful advice and consultation on special problems in mathematical logic and semiotics to Professor S. A. Yanovskaya first of all and then to the author's long-standing co-workers A. V. Kuznetsov, G. E. Kleynerman, V. K. Finn, and Docent Ye. K. Voyshvillo.

Contents

History of Mathematical Logic from Leibniz to Peano

CHAPTER ONE

The Development of Mathematical Logic During the Middle Ages in Europe

1. Survey and Periodization

Despite the fact that mathematical logic is primarily an offspring of the nineteenth century, individual ideas in this scientific discipline were brought to light much earlier. In fact, the foundations of a number of concepts in mathematical logic were laid in the Western European Scholastic tracts on logics.

The achievements of the logicians of the Middle Ages have not received enough attention in the Soviet literature. However, the foreign literature includes a number of important works. Source material on medieval logic is presented in a number of books, notably References 17 and 370. The works of K. Prantl in Reference 17 still command admiration for his scrupulousness (although we must remember that Prantl, who was not well versed in mathematical logic and wrote at a time when the subject was just beginning to develop as an independent scientific discipline, naïvely believed that such general logical tracts as *de consequentis* and *de insolubilis* were conglomerations of foolishness).

The Middle Ages has an important influence on the field of logic. The scholars of that time excelled in subtle and abstract arguments; however, isolation from experiment and the necessity of making their arguments agree in every detail with the tenets of Catholicism, the religion of many of the prominent researchers, were burdens on their creative powers. The existing atmosphere of verbal battles led to an increased interest in the study of the laws of deductive logic. In addition, this was sometimes used as a convenient means for escaping from the ideological domination of clerical dogmatism. Even Abelard, in the twelfth century, actually proved that the "theses of revelation" were inconsistent with the logical studies of the Greeks and Arabs. The fact that specific examples and religious content as well as "secular" illustrations were contained in logic is, of course, of no fundamental importance. Thus, as an example of a fixed true statement, the Scholastics frequently used the phrase *deus est* while, as an example of an impossible proposition, they used the phrase *deus non est*. Incidentally, the reader may find it humorous that such statements as *deus est bonus* (God is good) and *asinus est animal* (the ass is an animal) were taken to be logically equivalent.

The Scholastic logic of the Middle Ages can be divided into three phases. The first phase — the *vetus logica*, which included Aristotle's *On Commentaries*, his *Categories* with Porphyry's commentary, and the logical tracts of Boethius — extended into the middle of the twelfth century. The second phase, the *logica nova*, was characterized by the introduction, between 1136 and 1141, of the *Analytics*, *Topics*, and *On Sophistic Refutation* by Aristotle. The third phase began with the tract *On the Properties of Terms*, which led to the formation of the so-called modern logic (*logica modernorum*) in the form of various summulae.

Attempts to separate the development of medieval logic into periods are found in References 370 and 21. The first period extends from the beginning of the Middle Ages through the time of Abelard (roughly, through the middle of the twelfth century). The principal results of this period are related primarily to the works of Byzantine logicians, especially Michael Psellus (1018–1096), *Synopsis Organi Aristotelici Michaele Psello autore*, edited by M. Elia, Ehingero, 1597. Psellus was a Byzantine who lived in Constantinople. A platonist in his philosophic views, he subscribed to the Neoplatonic interpretation of the philosophy of Plato and Proclus. Of the latter Psellus said nothing without appending the epithet "the divine." However, the elements of Neoplatonism in the philosophical views of Psellus had no great effect on his work in logic. Psellus' main contribution to the Byzantine Renaissance was that, long before Spinoza, he applied the numerical methods of mathematics and geometry to proving philosophical statements.

The late Middle Ages tract *de terminorum proprietatibus* (on the properties of terms) has, for its starting point, a commentary of Psellus on the last chapter of Aristotle's *Topics*, which considers the importance of the various parts of speech from the viewpoint of logic.

According to Prantl, Western European scholarship became familiar with the *Synopsis* of Psellus in the thirteenth century (see *Synopsis quinque vocum Porphyrii et Aristotelis Praedicamentum*, Paris, 1541, and *Synopsis Logicae Aristotelis*, Augusta Vindelicorum [Augsburg], 1597, 1600).

In particular, Psellus examined the problem of the equivalence of propositions (*aequipollentia propositionum*) and found logical conditions required for such equivalence. Much space in the *Synopsis* is occupied by a study of substitutions of terms for other terms that would apparently be scarcely possible without explicit distinction between logical constants and logical variables. However, a large part of his logical research is nonetheless devoted to the problems that occupied the first commentators on Aristotelian logic during the time of Boethius and is associated primarily with a scrupulous study of various forms of syllogisms.

One of the most colorful figures in the first period of development of medieval logic was Peter Abelard (1079–1142). Among his works, published at various times, we should note *Introductio ad theologiam*, *Dialectica* (Reference 73), *De eodem ad diverso*, *Quaestiones naturales*, *Sic et non*, *A History of My Calamities* (St. Petersburg, 1902), *Die Glossen zu Porphyrius*, Beiträge zur Geschichte der Philosophie im Mittelalter (Münster, 1895).

In his commentaries on Porphyry, apparently written during his youth, Abelard went far beyond a simple interpretation of the Greek text. He defined logic as the science of evaluating and distinguishing arguments on the basis of their truth. Logic, said Abelard, teaches us neither how to use arguments nor how to construct them. Since arguments consist of propositions, and propositions consist of expressions, logic must begin with a study of simple statements and then more complex ones. Abelard devoted special attention to the copula and subjected it to logical analysis.

The problem of universals occupies an important place in the logical works of Abelard (this is, the problem of the nature of general concepts, which must be solved if a satisfactory theory of abstraction is to be constructed). The nature of the general, he said, depends not on words (*voces*) but on their meanings (*sermones*). His conception, which differed from that of more extreme nominalism, is sometimes called sermonism. Abelard attempted, first of all, to emphasize the semantic aspect of universals, that is, the relation between the symbols used for universals and the objects they represent.

Abelard's *Dialectics* contains an early form of a modally implicative treatment of conditional statements. For example, consider the following rules, which concern the truth of an implication:

1. If the antecedent is true, so is the consequent;
2. If the antecedent is possible, so is the consequent;
3. If the consequent is false, so is the antecedent;
4. If the consequent is impossible, so is the antecedent.

Abelard sometimes stated the first of these rules in the following form (Reference 73, p. 287):

1*. Nothing that is true ever implies something that is false.

According to Reference 50, p. 221, in one place Abelard used 1* as a general principle for construction of proper arguments.

The logical problematics of Boethius has a rather strong influence in a number of places in *Dialectics*. This is evident particularly in Abelard's discussion of equivalent expression of certain logical constants in terms of others. Thus, following Boethius, Abelard used the following equivalents:

5. $$(A \oplus B) \equiv \overline{(A \rightarrow B) \mathbin{\&} (B \rightarrow A)},$$

where $\oplus$ denotes the exclusive "or" (*Aut est A aut est B*). Abelard was also familiar with the use of disjunction in the ordinary (as we would now say, non-Boolean) sense (Reference 50, p. 223; *A vel B*). In particular he used the conjunction *vel* to combine predicates into categorical propositions. Incidentally, Abelard never chose definitely between the "exclusive" (Boethius) and "inclusive" (Chrysippus) use of the word "or." In the last analysis, he was inclined to treat disjunctive propositions as necessary disjunctions of the form $N(A \vee B)$, interpreting them as an alternative form of necessary implications of the form $N(\bar{A} \rightarrow B)$, where N is the necessity functor, and the antecedent of the implication is therefore negative (Reference 73, p. 489–499). Thus, Abelard placed the necessity operator before the weakened disjunction as a whole. For him, modal disjunction was in some sense more elementary than implication and other logical unions.

Another important logician of the twelfth century was English-born Adam of Balsham (*Parvipontanus* or Adam du Petit Pont), who taught the "trivium"* in Paris in his own school not far from Petit-Pont. Methodologically, he was opposed to the noted representative of early French Scholasticism Gilbert de La Porrée, (*ca.* 1070–1158/9). As a logician, Adam was known primarily for his adaptation of the "First Analytics" of Aristotle. The French historian Cousin refers to one of Adam of Balsham's manuscripts under the heading *Anno MCXXXII ab incarnatione Domini editus liber Adam de arte dialectica.*†

The Italian historian L. Minio-Paluello has analyzed the logical works of Adam in detail (Reference 74). He believes that Adam took a big step forward in his logic, in comparison with Aristotelian logic. For example, he conducted a detailed analysis of the liar's paradox in the form of the statement of a man claiming to be lying (*qui se mentiri dicit*).

As noted in Reference 79, Adam admitted the possibility of the existence of a set of objects with a proper subset having as many objects as the set itself. In other words, he anticipated Cantor's definition of an infinite class. Thomas Ivo (Reference 79) was therefore absolutely right in saying that one of the paradoxes of the infinite had already been stated in the twelfth century. In another place in his *Ars Disserendi*, the criterion Adam used to distinguish between finite classes and infinite classes was that the former do not have the property he noted as characteristic of the latter. But this actually anticipates C. S. Peirce's (nineteenth century) idea that a finite set cannot be mapped one-to-one onto a proper subset.

* The trivium was composed of grammar, rhetoric, and dialectics.

† Cousin, *Fragment d. philos. du moyen-âge*, Paris, 1885, p. 335. Concerning Adam of Balsham, see also Reference 17, vol. 2, pp. 104, 212, 213.

From the standpoint of mathematical logic, the period we have been considering is distinguished primarily by the attempts of Psellus to develop general methods for finding the "middle term" in connection with problems on finding propositions (*inventio propositionum*) and by his interest in logical symbolism, mnemonics, and finally a number of problems he considered on the logic of modality.

The beginning of the second period dates to approximately the middle of the twelfth century and continues until the last decade of the thirteenth. The most important logicians of this period were William Sherwood* (William of Shyreswood, died 1249), Albert von Bollstädt (Albertus Magnus or Albert the Great, 1193–1280), and their students Peter of Spain (1210–1277), John Duns Scotus (1270–1308), and Raymond Lully (1234–1315).

William was born in Durham in the last decade of the twelfth century and taught first at Oxford and later in Paris. He translated the *Synopsis* of Psellus long before Peter of Spain did. The *Synopsis* was the basis of the handbook of logic that Sherwood wrote in Latin; this book remained in manuscript form until it was rediscovered during the nineteenth century (*Codex Sorbonnicus*, 1797).

Sherwood was concerned primarily with a series of problems in modal logic. His study of the "liar's antinomy" became the basis for studies associated with analysis of undecidable propositions. This group of problems subsequently led to Middle Ages tracts on the theme *de insolubulis*. In addition, William devoted much attention to further development of the mnemonic methods of Psellus, among which we should note, for example, the well-known "logical square."

Peter of Spain's *Summulae Logicales* enjoyed great popularity as a text for teaching logic. Peter expended much effort in developing mnemonics to fit the needs of logic. His name is also associated with the revival of the traditions of the Peripatetic school in the Middle Ages. The logical works of Peter of Spain may be found in the following publications: *Textus omnium tractatum Petri Hispani*

* It appears that the source of William's surname was the same Sherwood Forest associated with the legendary Robin Hood.

(Venice, 1487, 1489, 1503; Cologne, 1489, 1494, 1496); *Summulae Logicales of Peter of Spain* (Notre Dame, Indiana, 1945).*

The *Summulae* consist of six sections: studies on suppositions, propagation, designation, distribution, limitation, and statements requiring additional interpretation. The *Summulae* present various mnemonic methods for remembering the rules of logic. Consider, for example, the following mnemonic verse for remembering the rules for proving equivalence of statements:

> *Non omnis, quidam non. Omnis non quasi nullus.*
> *Nonnullus, quidam: sed nullus non valet omnis.*
> *Non aliquis nullus. Non quidam non valet omnis.*
> *Non alter, neuter. Neuter non praestat uterque.*

It is easy to see, for example, that the first phrase of the first line (*Non omnis, quidam non*: "Not for all true, therefore false for some") corresponds to the formula $\overline{\forall x \mathfrak{A}(x)} \equiv \exists x \overline{\mathfrak{A}}(x)$, while the first phrase of the second line (*Nonnullus, quidam*: "Not for all false, so true for some") corresponds, in modern symbolic notation, to the relation $\overline{\forall x \overline{\mathfrak{A}}(x)} \equiv \exists x \mathfrak{A}(x)$, where the bar indicates the logical operation of negation, $\forall$ is the universal quantifier, $\exists$ is the existential quantifier, and $\equiv$ denotes equivalence. In the *Summulae* much attention is devoted to the tract *de terminorum proprietatibus* (on the properties of terms), which, in particular, analyzes words that in ordinary language have the same function as quantifiers have in formal logic, that is, the words every (*omnis*)†, no (*nullus*), both (*uterque*), neither (*neuter*), all (*totus*), and so on. There is also a special study of the construction of statements made by placing the word "only" before the subject of a sentence and of the elimination of statements containing the word "except" or containing suppositions, etc.

This last concept deserves further comment. The very word "supposition" literally means "proposition." The theory of suppositions is a theory of notation (for terms). A Scholastic would

* The literature on Peter of Spain includes References 25 and 61.

† Later Duns Scotus used the term *omnis* in the sense of the universal quantifier for predicates and the term *unusquisque* for the universal quantifier for individuals.

say that a term denoting a really existing object is being used in the mode of formal supposition (*suppositio formalis*). For example, in the proposition "the Earth is round" the term "Earth" denotes an actually existing object, the terrestrial sphere. On the other hand, a Scholastic would say that a term signifying a word that is the name of an existing object is used in the mode of material supposition (*suppositio materialis*). For example, in the proposition "'earth' consists of five letters," the term "earth" signifies a word (and not an object), and so is used in the mode of *suppositio materialis*. Although it is impossible to establish a direct connection between the suppositions of logicians of the Middle Ages and the precise logical notions of the present, there is, for example, reason to attempt to find semantic analogs for the use of terms in the mode of *suppositio materialis*.* The great popularity of the *Summulae* is indicated by the fact that it underwent forty-eight printings in the half century following the invention of the printing press.

Albert the Great (Albertus Magnus), a contemporary of Peter of Spain's, was born in Lauingen in Swabia, studied at Padua, and taught in Cologne and Paris. In the theory of abstractions, he leaned toward the compromise solution of the problem of universals, which was due to Avicenna. His writings were collected in *Opera Omnia*, vols. 1–38, Paris, 1890–1899.

Albert, who was strongly influenced by the Arabian Averroës (Reference 17, vol. 2, p. 105), was, in particular, responsible for development of Psellus' theory on methods of finding propositions; his primary goal was to solve the type of problem that appears when corollaries of given premises are sought. The problem he solved could be applied only to the syllogistic apparatus developed by Aristotle; hence his results were of a very special nature.

Particularly unique logical views were held by John Duns Scotus, who was also an outstanding political figure in the Middle Ages: he raised the problem of "apostolic poverty" to the hated papal curia, attacked the richness of the church, and exposed papal robbery.

Early in his career, Duns Scotus showed a great interest in

* Partially anticipating Peter of Spain, William Sherwood had already distinguished between formal and material supposition.

mathematical disciplines. At 23, he became a teacher at Oxford. Since he was a member of the Franciscan order, Duns Scotus was in outright opposition to the prevailing Catholic dogmatism of his day and, in particular, to the theological determinism of Thomas Aquinas. Duns Scotus' output is striking in scope. For example, the Lyon edition of his works (1639) consists of twelve volumes. Detailed analyses of the logical views of Duns Scotus may be found in Reference 56 and in E. Gilson, *Jean Duns Scotus*, 1952.

Following al-Farabi, Duns Scotus distinguished between the logic of the theoretical (*docens*) (as the science of necessary conclusions from necessary premises) and the logic of the applied (*utens*). He provided great impetus to the development of Peripatetic logic, basing his work primarily on the compendium of Peter of Spain. Duns Scotus, in particular, is responsible for the logical law which is formally expressed as: $p \supset (\bar{p} \supset q)$, where the bar indicates negation.

To a considerable extent, his ideas are responsible for intensive development of an important aspect of medieval logic, namely treatises on the theme *de obligatoris*. As Vladislavlev (Reference 19) notes, the formal impetus for these treatises is a single statement of Aristotle's (which occurs in his work twice — in the first book of the second *Analytics* and in the ninth chapter of *Metaphysics*): A premise accepted as really possible cannot entail an impossible conclusion. The Scholastics noted that if certain propositions are related (for example, one proposition implies another), then other propositions, on the contrary, can be treated as independent of one another.

One of the aims of the *de obligatoris* treatises consisted in finding logical conditions under which the assertion or the negation of some proposition belongs to this class of necessary implication. The notion of *obligatio* is defined as a defensible system of assertions such that none imply anything impossible, that is, in modern terms, the system of assertions satisfies the criterion of consistency.*

It appears that tracts on the theme *de obligatoris* can be treated, although far from absolutely so, as predecessors of axiomatic-

* "*Obligatio est praefixio alicjus enunciabilis ne sequitur impossibile*" (Reference 19).

deductive studies. Analysis of the logical legacy of Duns Scotus shows that it can be treated as a predecessor of not only the calculus of propositions in contemporary mathematical logic but also studies of Husserl type. It is scarcely necessary to add that the first tendencies in the logic and methodology of this prominent Scottish thinker were progressive.

Another bright figure in medieval logic was Raymond Lully. Although a number of authors have attempted to place him off the "main line" in the development of logic, it seems natural to say that the achievements of Lully (part of which, we should note in passing, belong to *de consequentis*) are a natural consequence of the raised level of logical formalization that was characteristic of the period after the *Summulae* appeared. It should be noted that Vladislavlev (Reference 18) attempted to overcome the nihilistic view of Lully and his school. According to Vladislavlev, Lully's "art" is a natural result of the evolution of logic in the Middle Ages.

Raymund Lully was born in 1234 in Palma, on the island of Majorca. Until he was nearly thirty, he was a courtier at the court of Jacob, King of Aragon, and he obtained some measure of acclaim as a poet. Lully's courtly life was conducted in very grand style. Once, in courting a married woman, he rode a horse at top speed into a church during a mass. As the legend has it (Reference 63), the "lady of his heart" said to her suitor: "Would you see my breasts, which your sonnets have so lavishly adorned? Well then, I can afford you that pleasure!" With these words, she lifted her mantle and exposed her breasts, which were covered with bleeding ulcers. This sight stunned Lully, and on the spot, he decided to devote himself to the church. The shock experienced by Lully is comparable to that experienced by Prosper Merimée's Don Juan in *A Soul in Hell* when, learning from a funeral procession that the "object of his affections" has died, Don Juan rapidly converts from a thoroughgoing hedonist to a hermit who confines himself to a monastic cell. The decision so quickly made by Lully abruptly changed his life and he removed himself from worldly vanity. Fortunately for science, Lully decided to devote his time to much more useful pursuits than the Don Juan of Prosper Merimée.

Exchanging the costume of a courtier for the habit of a monk, Lully devoted himself almost completely to the study of logic. He defined logic as the "art and science by means of which truth and untruth can be recognized by reason and separated from each other — the science of finding truth and eliminating falseness" (*Logica est ars et scientia, quae verum et falsum ratiocinando cognoscuantur et unum ab altero discernitur, verum eligendo et falsum dimittendo*); cited in Reference 17, vol. 3, p. 150. Lully's works are collected in *Raymundi Lulli Opera*, Argentorati, 1617. Of material on the nature and extent of Lully's logical ideas, the following sources should be emphasized: References 14, 15, 60, and 62.

After he developed the "great art," Lully threw himself into an ascetic life, attempting to find methods for practical application of his studies. He traveled to all the principal cities of Western Europe, acquainting the academic world with his discovery and simultaneously igniting a zeal among the clergy to eradicate Mohammedanism. He traveled to Africa three times, attempting to lead the "unfaithful" to the "way of truth." Lully spent the remainder of his life attempting, first, to build his own school of logic and, second, to "persuade" Mohammedans to change their relation to Christianity. He was successful in neither of his aims: a Lullist school of logic appeared only long after his death, and the Mohammedans held to their own convictions.

The device invented by Lully for mechanization of syllogistic processes was rather primitive and limited in its capabilities. Lully began by constructing four special figures. The *first figure* included nine absolute predicates ("is a quantity," "is perfect," etc.), nine subjects ("quantity," "perfection," etc.), and nine letters: *B*, *C*, *D*, *E*, *F*, *G*, *H*, *J*, and *K*. The outer ring of the associated circle was comprised of letters; then came subjects and, finally, predicates. The inner ring consisted of a circumscribed disk with a star-shaped figure drawn on it; the lines of this figure indicated the directions of all possible combinations of the initial concepts.

On the *second figure* nine relative predicates were written ("is different from," "is in the relation of agreement with," "is in the relation of mutual exclusion to," etc.).

Absolute predicates and relative predicates were combined in the *third figure*; combinations of the form *BB* were eliminated, and a combination of the form *XY* was assumed to be identical to the combination *YX*. The third figure generated two-term combinations of terms, that is, sentences.

The *fourth figure* was a system of three disks of which only the outermost was stationary. This one, according to Lully, could be used to simulate the syllogistic process, by which he meant the procedure of combining (and substituting) terms. Here the conjunction *est* was taken by Lully to express the relation between the whole and a part. It is interesting that, although Lully sometimes used the operation or quantification (refinement of volume) in an implicit form, he was, on the whole, far from a systematic analysis of this operation. In effect, the only operation in his logic of terms is the operation of intersection. This naturally greatly restricted Lully's combinatorial techniques and made it impossible to construct algorithms for his individual methods. For the structural details of Lully's figures, see Reference 19, pp. 100–107.

Among the members of Lully's school, we should note the following: the pantheist Giordano Bruno, Agrippa von Nettesheim (1487–1535), Athanasius Kircher (died 1680), J. H. Alsted (1588–1638), and the materialist and Epicurean Pierre Gassendi (1592–1655).

In his philosophical works, Lully fought Averroism and, in particular, the study of the dual nature of truth. Unfortunately, his arguments, as we would now say, were a "critique of the right."

In addition to his attempts to mechanically simulate logical operations, Lully also busied himself with analysis of the relations between the logical constants "and" and "or" as well as with a study of the logical essence of interrogative propositions. For example, in Lully's *Introduction to Dialectics*, statements are initially divided into true and false (Reference 19, p. 111). On pp. 151–152 of the same work, Lully presents conditions under which conjunctive and disjunctive statements may be true or false. These conditions correspond exactly to the modern rules for truth of conjunctions and disjunctions. And Lully's rule that "from the universal we can proceed to the corresponding particular, indefinite, and individual"

(*Introduction to Dialectics*, F8 rB) corresponds to the following three formulas of the calculus of predicates:

$$\forall x \mathfrak{A}(x) \vdash \mathfrak{A}(x_p) \text{ (to the particular!)}, \quad (1)$$

$$\forall x \mathfrak{A}(x) \vdash \mathfrak{A}(y) \text{ (to the indefinite!)}, \quad (2)$$

$$\forall x \mathfrak{A}(x) \vdash \mathfrak{A}(a) \text{ (to the individual!)}, \quad (3)$$

where $\forall$ is the universal quantifier and $\vdash$ denotes derivability.

The bridge between Lully's logic and that of the Jansenists of Port Royale was Lully's analysis of so-called general points of view (aspects) for study of various objects. The *loci communes* of the scholars of Port Royale was nothing more than a further development of Lully's "general points of view," directed toward a known standardization of problems on a certain theme (see Lully's work *Tractatus de venatione medii inter subjectum et praedicatum*). To the section on statements, Lully attempted to append the Scholastic theory of suppositions, which he greatly simplified: he separated suppositions into formal (simple and complex) and material.*

A particularly important achievement of the second period was the appearance of the *tractatus syncategorematibus*, which contains an analysis of simple and complex terms pertaining to the formal structure of propositions, for example, "not," "and," "if . . ., then," "each," etc. In the fourteenth century, the Scholastics augmented this treatise with a section entitled *Sophismata*, which

* Lully's simplification of the theory of suppositions was a direct consequence of well-known weaknesses of this viewpoint (for example, in the form used by Peter of Spain).

The theory of supposition was later criticized and even parodied. Francois Rabelais, for one, parodied it in the following manner in *Gargantua and Pantagruel* (Chapter XX, Book I):

As a sign of gratitude, Gargantua offers the Scholastic Janotus a bolt of cloth. Then there is an argument between Janotus and his colleagues, who are experts in logic, about who should get the cloth. "What is the supposition of this cloth?" (*Pannus pro quo supponit?*) asks Janotus. "Mixed and distributive" (*confuse et distributive*), says Bandouille "I am not asking you, blockhead, about the nature of this supposition (*quomodo supponit*) but about its object (*pro quo*). Answer: This piece of cloth was intended for my legs (*pro tibiis meis*), as the supposition carries the supplementaries (*sicut suppositum portat adpositum*)."

considers the most typical errors that appear in arguments, as well as an entire class of problematical statements.

The culmination of the achievements of the medieval logicians occurs in the third period in the development of Scholastic logic, which covers the epoch between William of Ockham (1300–1347) and the end of the Middle Ages (Ockham studied logic intensively between 1328 and 1329; see Reference 370). During this period a significant shift in Scholastic methodology occurred. In the first place, we should note the studies of intensification and remission of forms. Study of the variability of the intensity symbols other than the uniquely defined symbols of the Scholastics was a step toward the mathematical theory of variables (Reference 29).

This study is closely related to the attempt at symbolization in philosophy, which Maier characterizes as an "innovation of the fourteenth century"; she also considers the "inclination to mathematization of philosophical conceptions and proofs" (Reference 29, p. 79). Nonetheless, this tendency cannot be said to be fundamentally "new"; in essence, it is only a development of the older methodological ideas of the Chartres school.

The fourteenth century exhibits a distinctly new approach to the problem of the infinite; this approach is associated with abandonment of the "physicalization" of this notion by Aristotle. At this time, a considerably more abstract (than Aristotle's) base led to animated discussions of the relations between "categorematic" and "syncategorematic" infinity, which, to some extent, correspond to the modern problem on the relations between actual and potential infinity. Naturally, this controversy had an important influence on the process of refining the terms in logical works on the theme *tractatus syncategorematibus*. In these works, in addition to purely logical constants, there begin to appear analyses of such nonlogical notions as "infinitely many" and "infinitely small."

In view of the new tendencies of the fourteenth century in methodology, a symptomatic and unique (in his generation) attempt was made by the nominalist and skeptic Nicholas of Autrecourt (born about 1300, died about 1350) to revive the atomism of Democritus. His opposition to the current of orthodox Catholic dogmatism is eloquently indicated by the papal curia's censure of Nicholas' works in 1346 and by the subsequent burning of his "heretical" books.

The most prominent philosopher and logician of the Middle Ages, William of Ockham, was born at the very end of the thirteenth century. He studied at Oxford, where he later taught until 1324. As the result of a charge of heresy, Ockham spent four years in prison at Avignon. Rescued from his ideological adversaries, he found refuge at the court of Louis IV of Bavaria. Decidedly at odds with the papacy in the controversy over separation of church and state, Ockham supported the view that church and state should be completely separated. He admitted the possibility that the Pope and ecumenical councils might deviate into heresy, that is, he was actually a distant predecessor of the Reformation.* His conception of the duality of truth was sociologically reflected in his battle for secularization of the state. Ockham died of plague in Munich in 1347.

Many historians have given the name "terminism" to Ockhamist nominalism in the field of logic since Ockham believed that the point of logic is analysis of signs. In addition, he believed that the scientific methods applicable to logic, rhetoric, and grammar were closely related. "Logic, rhetoric, and grammar," he said, "are really cognitive studies, and not speculative disciplines, since they actually control the intellect in its activity."† Literature on Ockham can be found in References 1–10. In Boehner's opinion (Reference 3), Ockham considered three truth values: "true," "false," and "indefinite." We will denote truth by the numeral 1, falseness by numeral 0, and indefiniteness by N. Consider the function $p \supset q$ (if p, then q), defined on the set $\{0, 1, N\}$. For $p \supset q$ we construct the truth table shown in Figure 1, which, according to Boehner, is the form it took in Ockham's studies.

* In the dispute between Pope Boniface VIII and Philip the Fair, Ockham took the side of the temporal sovereign. He sharply criticized the Pope and his retainers in his *Defensorium* (a pamphlet). It is interesting to note that Ockham believed that it is impossible to prove the existence of God.

† Ockham's logical works are *Summa totius logicae*, Oxford, 1675; *Summa logicae*. Part 1, published by P. Boehner, Louvain, 1951; *Quodlibeta septem*, Paris, 1487.

In our opinion, we are not actually dealing, in this case, with a multivalued logic in the strict sense of the word. Rather, because "indefiniteness" is considered in addition to "true" and "false," the situation reflects a well-known occurrence in the logical square of elementary logic (for example, in this square the proposition that a partially affirmative statement is "true" implies that the

p	q	$p \supset q$
1	1	1
0	1	1
N	1	1
1	0	0
0	0	1
N	0	N
1	N	N
0	N	1
N	N	N

Figure 1

corresponding generally affirmative statement is only "indeterminate"). As a result, Boehner's assertion of the existence of a trivalent logic in Ockham's work seems to be a modernization of the problem.

Signs, according to Ockham, can be applied in two ways: artificially (words) and naturally (thoughts). Knowledge is composed of signs that, instead of reflecting reality, denote it. Signs become terms when they enter into propositions, and terms interpreted by the intellect form concepts; the value of spoken or written terms is, to a great degree, conventional, while concepts, according to Ockham, have a natural significance. Universals, in his opinion, are only terms that are used in logic to denote a group of objects and relations, but in no case do they actually exist as divine essences. Ockham believed that universals are also related to the mental act of understanding, that is, obliquely, with the external world. The essential pattern of abstraction, he asserted, depends on the so-called similarity of real substances.

Like all nominalists of the Middle Ages, Ockham carefully distinguished so-called connotative names (from the grammatical viewpoint, adjectives and participles). Thus, for example, the adjective "triangular" not only applies to a concrete triangle but also indicates its property of "triangularity." In other words, the general notion of "triangular" contains not only "concrete objects" (such as equilateral triangles) but "abstract" objects (such as "triangularity"). On the other hand, Ockham also discussed so-called absolute signs (such as "Plato," "virtue") each of which denote either only concrete objects (in a particular case, material objects) or abstract objects.

Ockham classified the sciences as rational or real. He defined the distinction between these two groups as follows: "Real sciences are distinguished from rational sciences by the fact that the terms, that is, the part of known propositions, in real science indicate and denote real objects, while rational sciences deal with terms that indicate and denote other terms" (Ockham, *Expositio aurea super totam artem veterem*).

The fact that Ockham's gnosiology is absolutely consistent not only with deductive methodology but also with inductive methodology was not, of course, accidental. Inductive empiricism gradually developed despite the triumph of deductive habits and concepts. The Scholastic proponents of the inductive theory of abstraction and the inductive method were Roger Bacon (1214–1294) and his teacher Robert Grosseteste (1175–1253).

Robert Grosseteste was born in Strandbrook, England (Suffolk). He made his reputation as a master at Oxford and a commentator on the physical and logical works of Aristotle. Of his writings, we should note *Roberti Grosseteste epistolae* ed. H. R. Luard (1861), vol. 25, *Rerum Britannicarum medii aevi scriptores*. Of his original works, we should note *Summa in octo physicorum Aristotelis libros*, 1498. He is discussed in References 77 and 78. According to Grosseteste, physics studies form as filled by matter, while mathematics abstracts form from matter. As a concession to the spirit of his time, he begins with the thesis that knowledge of universals requires intuitive comprehension. However, recognizing the role of intuition in gnosiology and ontology, Robert was not alien to rationalistic tendencies when he passed to formulation of problems in scientific methodology proper. Using the Aristotelian concept of mathematics as a more accurate physics (*Metaphysics* 12. 3. 1077b–1078a), Grosseteste went considerably further than the Stagirite, asserting that mathematics must be the basis of all physical disciplines.

"Logical and metaphysical truths, as a result of their distance from the senses and as a result of their precision," he said, "escape the intellect which must view them, as it were, from a distance; hence, the fine distinctions are not recognized. It is here that consideration from a distance, together with blurring of fine distinctions, proves to be the source of certain errors. Similarly, the reliability of physical conclusions is reduced because of the changeability of natural substances. It is these three — that is, logic, metaphysics, and physics — that Aristotle calls rational (*rationales*) since the unreliability of our comprehension of them requires us to make greater use of argument and probability than in the sciences; these latter, of course, contain both science and proof, but not in the most rigorous sense of the words. However, mathematics is at once a science and a proof in the most rigorous and proper sense (*maxime et particulariter dicta*)" (cited in Reference 76, p. 111).

Elsewhere, Grosseteste asserts that "all sources of natural actions must be given directly by lines, angles, and figures" (Reference 79, p. 293), which explicitly contradicts Aristotle's physicalism. On the whole, the methodology of Grosseteste can be characterized as mathematical atomism.

Thus, according to Grosseteste, mathematics is the most exact and exemplary scientific discipline. However, he took only the first steps along the path of practical application of mathematical methods to analysis of empirical data. The ideas that he developed concerning the necessity of experimental investigation of nature were later adopted and made concrete by the Franciscan monk Roger Bacon.

Roger Bacon was born into rich nobility near Ilchester in Dorset. In 1225 he entered Oxford and, after finishing his studies, remained there to teach. In Paris, Bacon joined the Franciscan order. Upon his return to Oxford in 1252, he lectured on mathematics, physics, and foreign languages. In 1257 he was banished from the university as a "wizard and sorcerer," and between 1278 and 1292 Bacon was imprisoned by the ignorant monks as a "dangerous heretic." The major works of Bacon are *Opus majus* (published 1733); *Opus secundum; Opus minus; Opus tertium* (*ca.* 1260); *Essays*, ed. Little (Oxford, 1914); *Summulae Dialectices* (Oxford, 1940). Bacon's work in logic is discussed in References 43 and 80–82.

According to Bacon, there are "two methods of knowing: by proof and by experiment. Proof yields a solution to a problem, but does not provide us with certitude, since the correctness of the solution need not be confirmed by experiment" (Reference 80, p. 18).

Bacon decisively separated theology from science and philosophy. In contrast to the qualitative estimates used by Aristotle in discussions of the world, he took important steps in the direction of rigorous quantitative evaluations of being.

In the fourth part of his *Opus majus*, he defines mathematics as the "alphabet of philosophy." "Mathematics has had the misfortune to be unknown to the church fathers," said this pioneer of Western European empiricism with biting irony. Mathematics, he believed, must be treated as an important goal in any branch of scientific endeavor.

"Mathematics provides," said Bacon, "universal techniques . . . that can be applied to all sciences . . . (and) no science can be known without mathematics" (translated by O. V. Trakhtenberg, Reference 43, p. 164).

"The problem of logic," Bacon said, "is to construct arguments that will move the practical intellect toward goals of love, courage, and felicitous being" (*Opus majus*, 59). No doubt this definition of the problems of logic is, to some extent, a flight of poetic rhetoric.

He includes logic, which he treats as an auxiliary scientific discipline like grammar, in the study of method.

In addition to Ockham and Bacon, among the outstanding logicians in the third period of the development of Scholastic dialectics we should note John Buridan (1300–1358), Albert of Saxony (1316–1390), John of Cornubia (second half of the fourteenth century), Ralph Strode (second half of the fourteenth century, peak in 1370), Marsilius von Inghen (1330–1396), Nicollette Paulus of Venice (Paul of Venice) (died 1429), and Peter of Mantua (second half of the fourteenth century). The third period witnessed the apogee of nominalism (and its variant, terminism), which occurred with the establishment of a methodological basis for logical investigations that is actually close to materialism. Note that the Paris physicist and Ockhamist John Buridan was the teacher of Albert of Saxony; the most extreme position of terminism was taken by Peter of Mantua; among the students of Buridan was Marsilius von Inghen; among students of the Scotists were Ockham and Strode.

The achievements of this era are highly respected in the classical works of Marxism-Leninism. In his article "Debate on the Free Press" Marx said, "The twenty tremendous volumes of Duns Scotus are as stunning . . . as a Gothic edifice," and from them we obtain a "real sense of value."* In another place, the founder of scientific communism said "Materialism is the unmistakable offspring of Great Britain. Even her great Duns Scotus wondered whether 'matter can think.'"†

Let us briefly summarize the problems dealt with by the nominalist logicians. The main direction taken by their research appears in the treatises *de consequentiis* and *de insolubiliis*, considered to be important supplements not only to Aristotelian logic but also to the compendium of Peter of Spain, which provided a direction for work on syllogisms and the theory of suppositions. The source of work in *de consequentiis* in the Middle Ages was the third book of the "Topics," where Aristotle focused attention on the existence of

* K. Marx and F. Engels, Collected Works (Russian translation) (Moscow–Leningrad, 1928), vol. 1, p. 140.
† K. Marx and F. Engels, *op. cit.*, vol. 2, p. 142.

contingent conclusions in which the assertions about one subject make it possible to make assertions about other subjects of the same type (Reference 19, p. 39). While the Scholastics were able to draw the idea of formal implication from Aristotle, for the elements of the theory of material implication they turned to the works of the Arabian logicians Avicenna, al-Farabi, al-Ghazzali (Algazel), and Averroës (Reference 17, vol 3, p. 138).

The tract *de insolubiliis* was brought to the nominalists via William Sherwood. The subject was highly developed by Thomas Bradwardine* (1290–1349), Walter Burleigh (1273–1357), Buridan, Albert of Saxony, and others. Many examples of semantic antinomies can be found, for instance in the sixth part of *Perutilis Logica Magistri Alberti de Saxonia* (Venice, 1592).

The liar's paradox was analyzed by Ockham in the third part of the fifth tract of his *Summa Logicae*. A theory of formal implication was constructed by Strode† in his work *Consequentiae Strodi* (Venice, 1493). The exceptionally broad range of logical research and the tremendous dimensions (even on the scale of contemporary works) of the logical compendia are astounding. Bochenski notes in Reference 370 that, for example, Paul of Venice's‡ *Logica Magna*

* This British Scholastic is noteworthy, in particular, for his attempts to apply mathematical methods of research to justification of a rationalistic theological system.

† Ralph Strode was a logician and pedagogue who taught at Oxford in the second half of the fourteenth century. He traveled widely, visiting France, Germany, Italy, Syria, and Palestine. The logical works of Strode (among them, in particular, *Consequentiarum formulae*) were placed in the *Consequentiae* and *Obligationes* classifications at the end of the fifteenth and sixteenth centuries.

‡ Paul of Venice, the Averroist from Udino, began teaching at Oxford in 1390 and wrote extensive commentaries on all of the fundamental philosophical and logical works of Aristotle. Paul's logical works comprise perhaps the highest state of Scholastic logic. He died in Padua in 1429. In addition to his *Logica Magna*, among his best known works we should note *Dubia circa philosophiam* (1493), published in 1498 under the title *Quadratura*. This work presents, in particular, the following classification of semantic antinomies:

1. The consequent is simultaneously correct and incorrect;
2. the statement is simultaneously true and false;
3. contradictory notions about the same object;
4. statements of mutual exclusion (Reference 11, p. 636).

(fourteenth century) contained about 3,650,000 words, that is, (according to the most conservative calculations), about the same amount as is contained in five or six volumes of a modern encyclopedia.

Aristotle's *Organon* provided great impetus to further evolution in logical studies. Leibniz considered the invention of syllogistics a major accomplishment of the human intellect. Thus far, Aristotelian logic had preserved its scientific value. Oriented basically toward gnosiological, linguistic, and grammatical factors, it nevertheless influenced the Greek mathematicians contemporary with Aristotle. A number of prominent men (among them Weyl) believe that the Aristotelian theory of proof has definite points in common with the structure of Euclidean geometry (Reference 23, p. 35). The usual characterization of Aristotelian logic as a logic of monadic predicates (properties) is, in general, correct.

However, discussions of the relationship between Aristotelian formalism and other logical studies must not neglect the fact that even Leibniz believed it possible (in principle) to reduce all relations to properties, and here he proceeded as a follower of the Stagirite. The consensus of many contemporary logicians is that Aristotelian logic includes the theory of formal implication and the beginnings of the calculus of modalities. It is not an accident, therefore, that the elements of mathematical logic appeared in the Scholastic tracts at the end of the twelfth century, when the Scholastics began to work directly from Aristotle's *Organon* and not from the investigations of Porphyry of Tyre (232–304) or A. M. T. S. Boethius (480–526), as had been the case before.

We should note that the scholars of the Middle Ages who dealt with logic constantly emphasized that they were the students of Aristotle and carefully cited all the places in his works that stimulated them to research in new directions.* The study of

* Some logicians have erroneously believed that Scholastic logic led away from Aristotle. In particular, this is the belief of the French logician Charles Serrius, who berated the nominalist logicians for illegitimately substituting words for ideas. This argument makes absolutely no sense: we must realize that no one has yet succeeded in operating with ideas outside his verbal–material skeletons (Reference 24, pp. 58–59).

mathematical logic in Scholastic treatises was stimulated by two fundamental circumstances: first of all, the influence of direct acquaintance with the original logical research of Aristotle and, second, the strengthening of sensualistic tendencies and the attraction toward problems in mathematics and natural science that were characteristic of the latter days of orthodox Scholasticism.

We will limit the analysis primarily to the characteristics of the two principal treatises (*de consequentiis* and *de insolubiliis*) that are of the most interest from the viewpoint of the history of logic (especially mathematical logic). We will first briefly describe the medieval logic of modalities and the *tractatus exponibilium* of Peter of Spain. Reference will not be made formally to the original texts except when it proves necessary for clarification of important points or when the meaning of the text is not clear. For literature on medieval logic, see References 1–63.

2. The Modal Logic of the Middle Ages

No special treatises on modalities were written during the Middle Ages, although discussions of modality appeared in various, more general works. Sherwood had already stated six forms of modal "modes": true, false, possible, impossible, contingent, and necessary. Later, Scholastics tended to limit themselves to three modes which, in their opinion, were of the most value: necessity, possibility, and impossibility. As the result of research on semantic antinomies, the logicians of the fifteenth century dealt primarily with the modes of truth, falseness, and undecidability (*insolubile*).

Sherwood was the first to apply the method of the so-called logical square (introduced by Psellus, see Reference 25) to analysis of the relations between statements that were related "materially" (i.e., in content) but were of different modalities. Sherwood's square of modal statements was later reproduced by his student Peter of Spain.

We denote the necessity functor by N, the possibility functor by M, the impossibility functor by U, and the contingent functor by T. Then Sherwood's square takes the form shown in Figure 2.

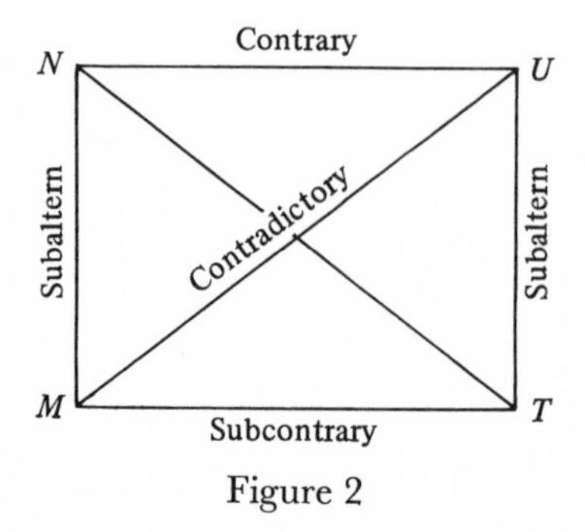

Figure 2

If, for example, N is true, then M is also true, T is false (N and T are in contradictory relation to each other), and U is false (N and U are in the contrary relation to each other). The same method of logical squares was applied to the study of direct modal conclusions by Ockham (Reference 17, vol. 3, pp. 317–318). We will consider, for example, one of his squares of modalities. Let N and M, respectively, be the modal necessity and possibility functors, where the symbols a, e, i, and o are respectively used to denote universal affirmation (XaY), universal negation (XeY), partial affirmation (XiY), and partial negation (XoY) in Aristotelian syllogistics. Ockham's square took the form shown in Figure 3.

The diagram in Figure 3 can easily be justified if we recall the relationship between the functors N and M, that is, $N(t) = \overline{M(\bar{t})}$, where the bar denotes logical negation and the symbol t denotes an

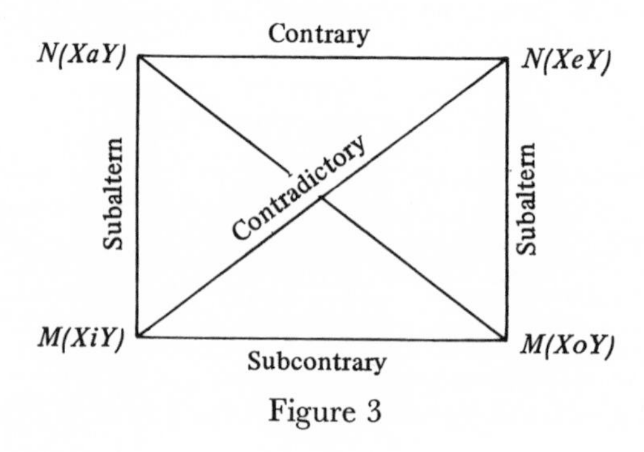

Figure 3

arbitrary statement. Then, if we substitute the expression XaY for t, we find that

$$N(XaY) = \overline{M\overline{(XaY)}} = \overline{M(XoY)},$$

and, by setting t equal to XeY, we find that

$$N(XeY) = \overline{M\overline{(XeY)}} = \overline{M(XiY)}.$$

In other squares, Ockham placed modal functors not only before a sentence but also before its components (that is, before both the subject of the sentence and its predicate). It should be emphasized that modal logic of propositions was, as a rule, still not differentiated from modal syllogistics, and the latter developed much more rapidly than the former. A special place was always occupied by the study of direct modal implications, which achieved its apogee in the work of Buridan.* Desiring mnemonic and graphical solidifications of the results he obtained, he was forced to abandon Psellus' square and to adopt a star-shaped figure (see Fig. 4). He considered modal functors (necessary, impossible, and possible) and the syllogistic functors a, e, i, and o. The subject–predicate structure of sentences was preserved, although modal functors as well as the symbol for logical negation could be placed before a statement as a whole or before any of the statement's individual components.

In Figure 4 we have substituted m for a and t for e to prevent confusion with the corresponding syllogistic functors. At each of

* John Buridan (1300–1358) studied logic, ethics, and physics. He taught at the University of Paris. His philosophical viewpoint comes under the heading of Ockhamism. From some time during the fifth decade through some time during the sixth decade of the fourteenth century, Buridan remained at the university where he was twice elected rector. Of his works, we should note the following: *Sophismata* (Paris, 1493); *Quaestiones super libros quattuor de caelo e mundo* (Cambridge, Mass., 1942); *Summa de dialectica* (Paris, 1487); *Perutile compendium totius logicae, cum jo. Dorp. expositione* (Venice, 1490). Buridan was known as a student of Ockham and was an intellectual determinist: In choosing one of several possible decisions, the "will" is controlled by reason. It acts only when the reason decides that one of the possibilities is the best (the will also chooses this possibility). But if reason decides that several possibilities are equivalent, the will does not act. This is the source of the well-known example of "Buridan's ass," which died while standing between two identical handfuls of grass.

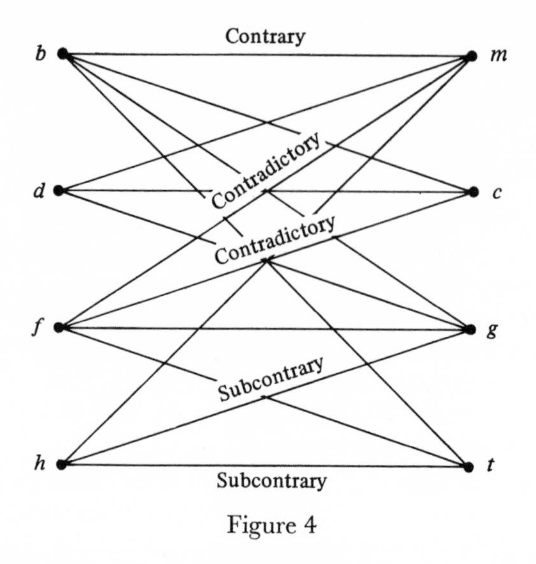

Figure 4

the vertices of the star-shaped figure, labeled m, b, c, d, g, f, h, and t, there is one of nine equivalent statements. Thus, it is possible to construct eight sets of statements (with nine in each set, where all statements in a given set are equivalent).

The following types of relations hold between these sets: contradiction, contrariness, inclusion (subordination), etc. We should note, for example, the contradictions. They constitute the following combinations: gb, hm, tb, fm, gd, and fc. Thus, for example, the sentence $Xa(N(Y))$, where N is the necessity functor and X and Y are the subject and predicate, respectively, is true at the vertex m if, let us say, the statement $Xo(N(Y))$ at the vertex h is false; conversely, $Xa(N(Y))$ is false if $Xo(N(Y))$ is true.

We will now present an example of equivalence of sentences of the set at the vertex m. The sentences $Xa(N(Y))$ and $Xe(M(\overline{Y}))$ of this set are equivalent, or we have that

$$Xe(M(\overline{Y})) \equiv Xe\overline{(N(\overline{\overline{\overline{Y}}}))} \equiv Xe\overline{(N(Y))} \equiv Xa\overline{\overline{(N(Y))}} \equiv Xa(N(Y)),$$

where $\equiv$ denotes logical equivalence.

The above examples give some idea of medieval research on

modal logic. Other, more interesting modal concepts of the Middle Ages are closely related to the contents of treatises on the theme *de consequentiis*, which we will consider in Section 4.

3. Analysis of Separative and Exclusive Sentences

Medieval logical treatises on the theme of exponibilia (exponibilia are propositions that require interpretation) are of interest because the study of exponibilia led the Scholastics close to the so-called De Morgan's laws. These rules were not explicitly stated by the Scholastics, although they were definitely used implicitly.

For separative exponibilia (that is, sentences of the form "only X has the property Y"), Peter of Spain presented the following four rules in his *Summulae* (Reference 19):

1. If a separative sentence is affirmative, it is equivalent to some conjunctive sentence whose first term is some affirmative statement (coinciding in content with the initial exponibilium without the word "only"), while the second statement is negative and such that its predicate is negated relative to all that is not the subject of the initial exponibilium.
2. "Only S is P" implies "every P is S."
3. If the symbol of negation is placed before the exponibilium, that is, if the exponibilium is of the form "it is not true that only S is P," then it is equivalent to a sentence of the form "it is not true that all S are P or (*vel*) something other than S is P."
4. The statement "only S is not P" is equivalent to an affirmative conjunctive sentence of the form "S is not P and all other than S are P."

Rules 1 through 4 admit the following symbolic representations (the expression X_Δ denotes "only X"); (1′ corresponds to 1, 2′ corresponds to 2, etc.):

1′. $(X_\Delta aY) \equiv [(XaY) \,\&\, (Xe\overline{Y})]$, where "&" denotes "and";
2′. $(X_\Delta aY) \supset (YaX)$, where $\supset$ denotes "if . . ., then";
3′. $\overline{(X_\Delta aY)} \equiv [\overline{(XaY)} \vee (\overline{X}iY)] \equiv [XoY \vee \overline{X}iY]$, where $\vee$ denotes inclusive "or" (*vel*);
4′. $(X_\Delta eY) \equiv [(XeY) \,\&\, (\overline{X}aY)]$.

In Expression 3, De Morgan's law $x_1 \,\&\, x_2 \equiv \bar{x}_1 \vee \bar{x}_2$ is used implicitly. Similarly, De Morgan's laws are used for analysis of exclusive statements (that is, sentences of the form "any S_1 except S_2 has the predicate P").

The mnemonic aids of the logical square have been widely used for study of the relations between exponibilia that distinguish quantity and quality (that is, on the basis of the volume of the

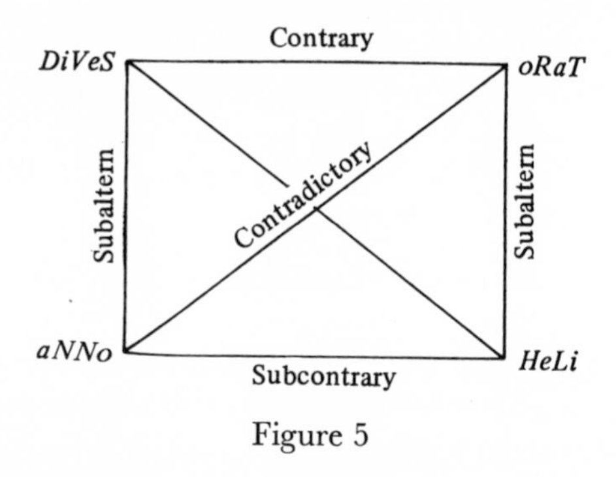

Figure 5

subject or the affirmative or negative nature of a copula) or that distinguish between the subject and predicate of a statement (in the corresponding sentence).

Consider, for example, the square constructed by Peter of Tartaret* (see Reference 17, vol. 4, p. 208). This square (or more properly, "metasquare" since it contains several logical squares) took the form shown in Figure 5.

As usual, *i*, *a*, *o*, and *e* are the syllogistic functors, while the letters *D*, *S*, *R*, *T*, *N*, *H*, *V*, and *L* have no semantic meaning — they are simply parts of the mnemonic signs.

Assume that we are dealing with exponibilia of type *i*, that is,

* Peter of Tartaret was a fifteenth-century Scotist and an active commentator on the logical, physical, and ethical works of Aristotle. A Franciscan who is representative of the later Scholastics, he became rector of the University of Paris in 1490. Of his works, we note *Expositio in Summulas* (1501); *In Summulas Petri Hispani, in Isagogen Porphyrii et Aristotelis Logicam* (Venice, 1592).

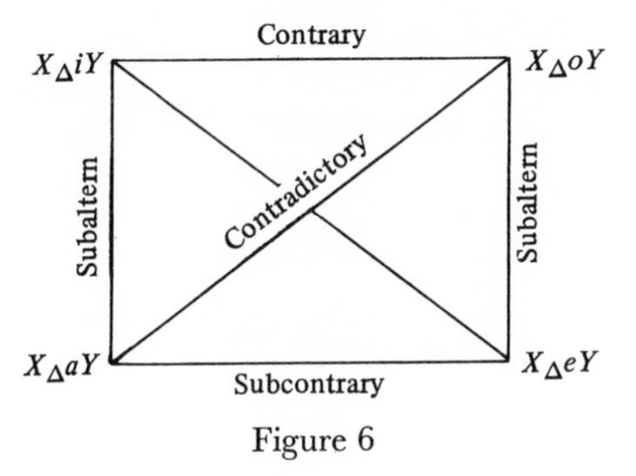

Figure 6

with expressions of the form $X_\Delta iY$. For this case the "metasquare" of exponibilia takes the form shown in Figure 6:

$$X_\Delta iY \equiv \overline{X_\Delta eY} \equiv \overline{XeY \,\&\, \bar{X}aY} \equiv (XiY) \vee (\bar{X}oY)$$

$$X_\Delta oY \equiv \overline{X_\Delta aY} \equiv \overline{XaY \,\&\, \bar{X}eY} \equiv (XoY) \vee (\bar{X}iY)$$

If $X_\Delta aY$ is true, then $X_\Delta oY$ is false.

In conclusion, we note some of the fundamental rules of negation for syllogistic propositions. These rules are explained in many places, so we will present them immediately in their symbolic form:

1. $\overline{SaP} \equiv SoP$;	2. $\overline{SeP} \equiv SiP$;	3. $\overline{SoP} \equiv SaP$;
4. $\overline{SiP} \equiv SeP$;	5. $Sa\bar{P} \equiv SeP$;	6. $Se\bar{P} \equiv SaP$;
7. $Si\bar{P} \equiv SoP$;	8. $So\bar{P} \equiv SiP$;	9. $\overline{Se\bar{P}} \equiv SoP$;
10. $\overline{Sa\bar{P}} \equiv SiP$;	11. $\overline{Si\bar{P}} \equiv SaP$;	12. $\overline{So\bar{P}} \equiv SeP$.

Despite the obvious use of De Morgan's laws in certain places in the *tractatus exponibilium*, it took a long time to find medieval works that would make it clear whether the Scholatics were aware of the meaning of De Morgan's laws in their general form (that is, without special conditions on the formal structure of the variables treated by the laws). Finally Boehner (Reference 26) succeeded in demonstrating that Ockham used De Morgan's laws. A number of logical historians (such as J. Lukasiewicz, Reference 27) believe that one could say Ockham's laws instead of De Morgan's laws. For the history of the application of De Morgan's laws during the Scholastic period, the interested reader is referred to the valuable material of Reference 26.

4. The Theory of Logical Implication

Here we will present a detailed analysis of the division *de consequentiis*. Fundamental to this section is the notion of implication, which is rather broad.

First of all, this notion includes the modern view stated by the expression "A implies B" (B can be derived from A by way of definite rules; symbolically, $A \vdash B$, where implication is meant in the sense of some relation between A and B).

Secondly, this notion includes the idea corresponding to interpretation of implication as an operation or function. In modern "language," this approach corresponds, for example, to the symbolic relation $A \supset B$, which, without stretching the point, can be read "If A, then B." Interpretation of implication as an operation has been traced to the Megarian-Stoic school (Philo). This approach, however, was comparatively rare in the logic of the Middle Ages, and, as we shall shortly see, it was unknown to Duns Scotus, who treated implication in the sense of derivability.

Duns Scotus classified the forms of implication as follows: "Implications may be separated into proper and improper. Proper implications are enthymematic or syllogistic. Enthymematic implications can be separated into formal and material, the latter being simple or factual."*

According to Duns Scotus, proper implication (*bona consequentia*) is concerned with a hypothetical relation in which the antecedent and the consequent are related by a contingent or causal relation such that it is impossible for the antecedent to be true and the consequent to be (simultaneously) false.† Duns Scotus also claimed

* Duns Scotus, *Qu. Sup. Anal. Pr*, 1, 20, p. 302 B; see Reference 17, vol. 3, p. 139.
† *Ibid*, 10, p, 287 B: "*Consequentia est propositio hypothetica composita ex antecedente et consequente mediante conjunctione conditionali vel rationali, quae denotat, quod impossibile est ipsis, quod antecedens sit verum et consequens falsum*." See Reference 17, vol. 3, p. 139, fn. 614.

that any proper implication that is not an Aristotelian syllogism is an enthymeme. He characterized "formal implication" as follows:

"A formal implication occurs when . . . both terms can be transformed (*uterque terminus est convertibilis*)" (Reference 17, vol. 3, p. 229, fn. 199).

Although it might seem from the above quotation that Duns Scotus included material implication in the modern sense in the notion of formal implication in a similarly modern sense, it is clear from elsewhere in his work that he used the terms *consequentia materialis* and *consequentia formalis* in other than the modern sense.

He associated the first term with conclusions that cannot be obtained in a purely formal manner and, if the conclusion is to be logically justifiable, must be supplemented by additional information. (The simplest case of such conclusions is the common enthymeme of elementary logic.) On the other hand, he associated *consequentia formalis* with conclusions of a purely formal nature. However, *consequentia materialis* can be reduced to *consequentia formalis* by reconstitution of omitted premises, after which the content of the implication plays absolutely no part in determining the logical correctness of the conclusion. Material implication can be reduced to formal implication in two different ways.

Duns Scotus called material implications that could be reduced to formal implications by appending a statement of necessity to the antecedent "simple material conclusions" (*consequentia materialis simpliciter*). For example, we can reduce the implication "Man walks. Therefore, animals walk," to a formal implication (in the Duns Scotus sense) by adding the statement "Every man is an animal."

On the other hand, Duns Scotus called a material implication that he reduced to a formal implication by appending a contingent statement to the antecedent "a factual (*ut nunc*) material implication." Thus, for example, the proposition "Socrates goes. Therefore, what is white goes," can be reduced by constructing the contingent statement "Socrates is white."

Although Duns Scotus treated the relation of logical implication as dealing with content, he and his closest student, William of

Ockham, treated implication itself in a manner very close to that of formalization provided by material implication. Thus, among Ockham's rules for implication, we might note, for example, the following: (1) An impossible statement implies any conclusion. (2) A necessary statement follows from whatever is convenient.*

These rules correspond to the following identically true expressions in the propositional calculus: $A \supset (\bar{A} \supset B)$ and $A \supset (B \supset A)$, where the bar denotes logical negation.

Elsewhere, Ockham states the following rules: (3) A true statement never implies a false statement. (4) A possible statement never implies an impossible statement (Reference 17, vol. 3, p. 129).

These rules correspond to the following formulas in propositional calculus and modal logic:

$$A \,\&\, \bar{B} \supset \overline{(A \supset B)} \quad \text{and} \quad ((A \equiv p) \,\&\, (B \equiv imp)) \supset \overline{(A \supset B)},$$

where p denotes some fixed statement of possibility (from the Latin *possible*) and *imp* denotes a fixed statement of impossibility (from the word *impossibile*).

We should also note certain other of Ockham's rules of implication (see Reference 17, vol. 3, pp. 390, 392, 396, 411–415, 417–419): (5) A false statement may imply a true statement (*ex falsis potest sequi verum*). (6) What follows from the consequent also follows from the antecedent (*Quidquid sequitur ad consequens, sequitur ad antecedens*). This is a statement of the factual transitivity of implication. (7) Necessity does not imply contingency (*Ex necessario non sequitur contigens*). (8) What contradicts the consequent contradicts the antecedent (*Quidquid repugnat consequenti, repugnat antecedenti*). We can formalize Statement 7 as follows:

$$(x \equiv N(x)) \,\&\, (y \equiv (M(y) \,\&\, M(\bar{y}))) \supset \overline{(x \supset y)},$$

where $N(x)$ denotes the statement "x must occur," and the conjunction $M(y) \,\&\, M(\bar{y})$ states that y is contingent (here contingency is defined as the possibility of M: M may or may not be true).

We should keep in mind that Ockham took necessity to be the

* Ockham, *Summa Logicae*, c. 70A, cited in Reference 17, vol. 3, pp. 129–130.

negation of contingency, so his seventh rule is actually only a restatement of his third (it is sufficient to replace the term "true" in the latter with the term "necessary" and the word "false" with the word "contingent"). Of course, it should be noted that here there is no assertion of metaphysical identification of the categories of truth and necessity on the one hand and the notions of falseness and contingency on the other; there is merely a statement that necessity bears the same relation to contingency as truth to falseness.

Duns Scotus' successor Ralph Strode developed what is actually a theory of formal implication (in the sense of Duns Scotus), for which he stated twenty-four rules (Reference 17, vol. 4, pp. 46–48, 50–52). Some of these rules are interesting because Strode introduced a further gradation in the modes of modal logic. He included the mode "questionable" between "true" and "false." Some of his rules read as follows: (9) If the consequent is questionable, the antecedent is either questionable or known to be false (*Si consequens est dubium, et antecedens est dubium vel scitum esse falsum*). (10) If the antecedent is questionable, the consequent is not necessarily negative (*Si antecedens est dubitandum, ergo consequens non est ab codem negandum*). (11) If the consequent is negated, the antecedent is not necessarily questionable (*Si consequens est negandum, antecedens non est ab codem dubitandum*). (12) If the antecedent is known, the consequent is known (*Si antecedens est scitum, et consequens est scitum*). This rule states perfectly the essentially analytic nature of implication in the sense of deductive implication. (13) If the consequent is impossible, the antecedent is also impossible (*Si consequens est impossibile, igitur et antecedens est impossibile*). (14) If the consequent is contingent, the antecedent is either contingent or impossible (*Si consequens est contigens, et antecedens est contingens vel impossibile*).

Analysis of other rules stated by Strode shows that, together with material implication* in the sense of the Stoics and Avicenna, and

* Strode defines material implication as follows: "Material implication is governed by two rules: (1) An impossible statement implies anything; (2) A necessary statement follows from anything" (*Pro consequentia materialis sunt duae regulae*: (1) *ex impossibile sequitur quodlibet*; (2) *Necessarium sequitur ad quodlibet*).

formal implication in the sense of Duns Scotus, he permitted a type of implication that was not considered by either the Stoics or Duns Scotus. This is made clear by Strode's remark, "a conditional relationship may be false, even if the antecedent and consequent are simultaneously true" (. . . *Conditionalem esse falsum, cuis tam antecedens, quam consequens esset verum*). Strode provides the following example of such a false conditional relationship: *Si tu est homo, ergo sum homo.*

Thus, Strode moved away from the Stoic notion of implication (which he did not reject entirely, for he reserved it for particular types of sentences) toward the notion of a type of implication which might be called semantic and which assumes that the consequent must be part of the antecedent if the corresponding conditional relationship is to be true. The material available does not permit us to determine which of the contemporary notions of implication — the Lewisonian or the Ackermannian (Reference 28) — Strode was moving toward; we can only be sure that he abandoned the treatment of implication as strictly a function of truth, an approach taken by the logical school of the Stoics.

The rules of implication stated by Albert of Saxony are also of interest (Reference 17, vol. 4, pp. 73–75). In particular, he stated the rule that, if A, together with some statement of necessity, implies B, then B follows from A alone.* Symbolically, we have—

$$(15) \quad \frac{A,\, N \vdash B}{A \vdash B},$$

where the N is associated with the phrase "statement of necessity" (from the word *necesse*) and the symbol $\vdash$ denotes derivability (following from the preceding).

Marsilius von Inghen (Reference 17, vol. 4, pp. 101–102) was responsible for an important contribution to the development of *de*

* Reference 17, vol. 4, p. 73: *Logica Alberticuii*, C2, F24rB: *Si ad A cum aliqua necessaria sibi apposita sequitur B, tunc B sequitur ad A solum.* In the literature, Albert of Saxony is also called Albert von Riegensdorf. He is known as a mathematician and commentator on the works of Aristotle and Ockham. After syllogistics, he considered the *Topics*, which he treated as part of a study on implication in the form of a theory of so-called maximal propositions (*propositione maxima*) (Reference 17, vol. 4, p. 78).

*consequentis.** Among the rules for which he is responsible we should note the following:

(16) The entire disjunction follows from each of its terms.†

(17) A universal statement makes it possible to conclude arbitrary individual terms.‡

It is easy to see that the sixteenth rule corresponds to the modern rules

$$A \vdash A \vee B, \quad B \vdash A \vee B,$$

for introducing the sign of disjunction, while the seventeenth rule corresponds to the rule

$$\forall x \mathfrak{A}(x) \vdash \mathfrak{A}(y)$$

of the restricted calculus of predicates, where $\forall$ denotes the universal quantifier and can be read "for all x." The letter $\mathfrak{A}$ denotes some property (predicate) of the subject x. The expression "the subject x has the property" can be symbolically written in the form $\mathfrak{A}(x)$.

In his work *Logica magistri Petri Mantuani* (published in 1483 in Pavia and in 1492 in Venice), the terminist Peter of Mantua considered the problem of *de consequentiis* in close connection with problems of modal logic. However, he is important not because of his results as a whole but because of one of the rules pertaining to the problem of the modality of rules, namely, the rule that leads to the conclusion *ad nullam de necessario*, that is, in contemporary terms, a necessary statement follows from the empty set of premises (*ad*

* Marsilius von Inghen was a German successor of John Buridan. He received his education in Paris, reached the level of magister, and became rector of Heidelberg. His *Dialectics* was appended in 1512 to an edition of Peter of Spain's *Summulae*.

Doctrinally Marsilius was a Thomist, but in the theory of material implication he leaned toward the views of Duns Scotus; in logic he was at one with Ockham. In the theory of abstraction he adopted a compromise position between realism and nominalism. Marsilius wrote glosses for Aristotle's *Categories* and for the well-known *Introduction* of Porphyry. Among his works we should note *Quaestiones super quator libros Sententiarum* (*Petri Lombardi*) (1947), which was published in a second edition in 1501.

† Reference 17, vol. 4, pp. 101–102.

‡ *Ibid.* Raymond Lully stated this rule in a somewhat more complicated form: "From a universal statement we can conclude the corresponding particular, indeterminate, and individual statements."

nullam). In addition to this rule, Peter stated sixty-three other rules of the same type.

Gradually a tendency arose to consider the general notion of logical implication. As far as we can judge from Prantl's classification (Reference 17, vol. 4, p. 181) of views on the nature of implication in medieval logic, there were three fundamental approaches to the definition of logical implication. According to the first view, implication is the relation of derivability of a consequent from an antecedent (Duns Scotus, Ferabrikh); according to the second, implication consists in establishing a situation where the consequent is part of the antecedent (Strode, Ockham); according to the third, implication is nothing more than an aggregate consisting of an antecedent, a consequent, and a conjunction (the extreme terminists). The first conception is close to the theory of formal implication, the second is very nearly the same as what is meant by implication in the semantic sense, and the third treats implication in a manner close to that admitting formalization in terms of the apparatus of material implication.

5. The Theory of Semantic Paradoxes

Paradoxes of the "liar's" type had been dealt with as early as the time of Adam of Balsham (twelfth century). Basically, however, rapid progress concerning *de insolubiliis* occurred during the fourteenth and fifteenth centuries. Important contributions to these studies were made by Albert of Saxony and John Buridan.

Buridan, in particular, considered such *insolubilia** (that is, paradoxical statements) as the statement "everything written in this volume is false," where this statement itself is the only thing written in the entire volume.

We denote the proposition in quotes by p. It is required to determine whether p is true or false. Assume that p is true. But then, as the statement says, it is false, because it is written in the given book. We now assume that p is false. But then the book should contain at

* Buridan's statement of this *insolubilium* reads as follows: *Propositio scripta in illo folio est falsa*. We have presented it in a form somewhat more convenient for analysis.

least one true proposition, and this can be only p since it is the only proposition in the given book. As a result, p must be true. Thus, if we assume that p is true, we must conclude that it is false, and conversely, if we assume that p is false, we must conclude that it is true.

As a result, we have before us two statements of the form $p \supset \bar{p}$ and $\bar{p} \supset p$, demonstrating the presence of a paradox since it follows from them that p is equivalent to $\bar{p}$, that is, either both statements are true (we have p and $\bar{p}$) or both are false (we have $\bar{p}$ and $\bar{\bar{p}}$). In both cases, we have an explicit formal contradiction since both the proposition and its negation are true (in the second case both are false: we have $\bar{p}$ and $\bar{\bar{p}}$).

We will now present several examples of the semantic antinomies encountered in the work of Albert of Saxony that permit, without changes in form, changes in some part of the content of the paradoxical statements within them. We state the first of these antinomies in the following form:

"If $2 \times 2 = 4$, then some contingent statement made by a liar in a given time interval t is false," and, except for the phrase in quotes, over the time interval t the liar N makes no contingent statements.* It is required to determine whether the liar N has made a true or false statement.

To rigorously analyze this paradoxical situation, we initially introduce two predicates: (1) during the time interval t someone has asserted x," which (predicate) we denote by "ass x," and (2) the identity predicate, which has the property that, if x and y are identical (that is, if we have $x = y$), then all that is true concerning x is also true concerning y. We can now state the problem in the form

1′. $\text{ass}(2 \times 2 = 4) \supset (\exists X \exists Y((\text{ass}(X \supset Y)) \,\&\, \overline{(X \supset Y)}))$,

which is a symbolic representation of the sentence "someone says, 'if $2 \times 2 = 4$, then some stated contingent proposition is false.'"

2′. $\forall t \forall v(\text{ass}(t \supset v)) \supset (t = (2 \times 2 = 4))$
$\& \, (v = \exists X \exists Y(\text{ass}(X \supset Y) \,\&\, \overline{(X \supset Y)}))$,

* *Si deus est, aliqua conditionalis est falsa, et sit nulla alia conditionalis* — Albert, Perutilis Logica Magistri Albert de Saxonia, VI (Venice, 1522). Albert set out to determine whether the proposition in italics is true or false (assuming that it is unique).

which states that for N any contingent statement made by N in the time interval t coincides with that of 1′. We now assume something that N said, that is,

$$2 \times 2 = 4 \supset \exists X \exists Y(\text{ass}(X \supset Y) \ \& \ (\overline{X \supset Y})) \tag{1}$$

is true. Since the premise of the implication is true, in view of our assumption the conclusion is also true, that is, the expression

$$\exists X \exists Y(\text{ass}(X \supset Y) \ \& \ (\overline{X \supset Y})) \tag{2}$$

is true.

Applying a common procedure of introducing symbols U_0 and V_0 to be treated as constants (that is, for which nothing can be substituted and to which we cannot apply the rule of generalization), we write Implication 2 in the form

$$\text{ass}(U_0 \supset V_0) \ \& \ (\overline{U_0 \supset V_0}). \tag{3}$$

Applying the rule for elimination of the universal quantifier (*dictum de omni*)* to Condition 2′, we find also that

$$\text{ass}(U_0 \supset V_0) \supset (U_0 = (2 \times 2 = 4)) \\ \& \ (V_0 = \exists x \exists y(\text{ass}(X \supset Y) \ \& \ (\overline{X \supset Y}))). \tag{4}$$

Since, in view of Implication 3, the premise $\text{ass}(U_0 \supset V_0)$ in Implication 4 is true, the consequent of Implication 4 is also true; that is, U_0 coincides with $2 \times 2 = 4$, and V_0 coincides with

$$\exists x \exists y(\text{ass}(X \supset Y) \ \& \ (\overline{X \supset Y})).$$

But this means that $U_0 \supset V_0$ is precisely our Implication 1. Nonetheless, Implication 3 implies that $\overline{U_0 \supset V_0}$. Assuming that Implication 1 is true (that is, that $U_0 \supset V_0$), we thus find that Implication 1 is false.

We now assume that Implication 1 is not true. Since its antecedent is true, the conclusion must be false; that is, we have

$$\overline{\exists x \exists y(\text{ass}(X \supset Y) \ \& \ (\overline{X \supset Y}))} \quad \text{or} \quad \forall x \forall y((\text{ass}(X \supset Y)) \supset (X \supset Y))$$

* We do this by substituting U_0 for t and V_0 for V, which this rule permits us to do.

from which it follows, by virtue of the rule for elimination of the universal quantifier by replacing X by $2 \times 2 = 4$ and Y by $\exists x \exists Y(\text{ass}(X \supset Y) \ \&\ (\overline{X \supset Y}))$, that

$$\text{ass}(2 \times 2 = 4) \supset \exists x \exists y(\text{ass}(X \supset Y) \ \&\ (\overline{X \supset Y})) \\ \supset (2 \times 2 = 4) \supset \exists x \exists y(\text{ass}(X \supset Y) \ \&\ (\overline{X \supset Y})). \quad (5)$$

Since by Condition 1′ the antecedent of Implication 5 is true, the conclusion is also true. But this conclusion is also the consequent of Implication 1, from which it follows that all of Implication 1 is true, even though we assumed it to be false. Thus, if Implication 1 is true, it is simultaneously false. Similarly, if it is false, it is also true. We are therefore dealing with an unsolvable proposition, a paradox.

We can analyze the following examples of Albert of Saxony (which have been changed in content but not in form) in the same way:

a. "$2 \times 2 = 5$ or some disjunction is false," where the universe of discourse contains no more than one disjunctive proposition;*

b. "$2 \times 2 = 4$ or some conjunction is false," where the conclusion in quotes is the only conjunction in the universe of discourse;†

c. "Socrates says, 'Man is an animal,' while Plato asserts that 'only Socrates speaks the truth,' and there are no other assertions. It is required to determine whether or not Plato has lied."‡

We will analyze Example c. First of all, it is quite easy to obtain a paradox contensively. Assume that Plato is telling the truth. In this case only the assertion "Man is an animal" is true, and all others are false.

But Plato's statement differs from Socrates'. Thus, Plato has lied. We now assume that Plato has lied. Then we must find at least one true statement other than that of Socrates. But the only other possible statement is Plato's (recall that, except for his statement,

* Albert de Saxonia, *Logica . . .*, VI; '*Homo est asinus vel alique disjunctiva est falsa,' et sit nulla alia disjunctiva in mundo.*

† *Ibid.*; '*Deus est et aliqua copulativa est falsa,' et sit sic, quod nulla alia copulative sit in mundo, haec ipsa. Tunc quaeritur, utrum sit vera.*

‡ *Ibid.*; *Dicat Socrates: "Deus est," et Plato dicat: "Solus Socrates dicit verum," et non simul alii loquentes in mundo. Tunc quaeritur, utrum Plato dicit verum.*

there is no other). As a result, Plato is not lying; so we have a paradox.

To formalize this paradox, we denote the statements of Socrates and Plato by p_1 and p_2, respectively; we introduce the identity predicate,* and we introduce the predicates SS (Socrates said) and PS (Plato said). The antinomy is formalized by the following system of premises:

1′. p_1 (in other words, p_1 is a fixed true statement).
2′. $SS(p_1)$.
3′. $p_2 \equiv SS(p_1) \,\&\, \forall z((z \neq p_1) \supset \bar{z})$.
4′. $PS(p_2)$.
5′. $\forall w((w = p_1) \vee (w = p_2))$, which means that except for p_1 and p_2, no other statement is made in the time interval t.
6′. $p_1 \neq p_2$.

We assume that what Plato said is true, that is, the following formula is true:

$$SS(p_1) \,\&\, \forall z((z \neq p_1) \supset \bar{z}). \tag{1}$$

But in Formula 1 the first conjunctive term is true (Premise 2′), it can therefore be eliminated. Formula 1 is equivalent to the expression

$$\forall z((z \neq p_1) \supset \bar{z}). \tag{2}$$

By *dictum de omni*, it follows from 2 that

$$(p_2 \neq p_1) \supset \bar{p}_2. \tag{3}$$

But because the antecedent of Implication 3 corresponds to Premise 6′, it follows that 3 is equivalent to

$$\bar{p}_2. \tag{4}$$

Thus, if we assume that p_2 is true, we are led to the conclusion that it is false.

Assume that p_2 is false, that is, the formula

$$\overline{SS(p_1) \,\&\, \forall z((z \neq p_1) \supset \bar{z})} \tag{5}$$

* It is unnecessary to introduce a special distinction predicate, because distinction is nothing more than simple negation of identity.

is true; this formula is equivalent to

$$SS(p_1) \vee \exists z((z \neq p_1) \ \& \ \bar{\bar{z}}), \tag{6}$$

from which it follows, by Premise 2′, that

$$\exists z(z \neq p_1) \ \& \ \bar{\bar{z}}. \tag{7}$$

We now introduce the symbol z_0 to replace z, using the same technique we already used in formalizing the preceding paradox. We have

$$(z_0 \neq p_1) \ \& \ \bar{\bar{z}}_0. \tag{8}$$

Applying the rule for elimination of the universal quantifier to Premise 5′, we find that

$$(z_0 = p_1) \vee (z_0 = p_2). \tag{9}$$

From 8, 9, and the fact that $\bar{\bar{z}}_0$ is equivalent to z_0, it follows that z_0 coincides with $\bar{p}_2$. By substituting p_2 for z_0 we obtain:

$$(p_2 \neq p_1) \ \& \ p_2. \tag{10}$$

But the first term of 10 can be eliminated since it is Premise 6′. Thus, the final statement is:

$$p_2. \tag{11}$$

As a result, if we assume that p_2 is false, we find that p_2 is true, a fact which in conjunction with Formula 4 leads to an antinomy. We leave it to the interested reader to obtain antinomies from the following two examples of Albert of Saxony, which have been slightly changed in content but not in form:

d. We are given three statements p_1, p_2, and p_3, where p_1 states that p_2 is false, p_2 states that p_3 is false, and p_3 states that p_1 is false. Determine whether or not p_1 is false.

e. We are given only the two statements p_1 and p_2, where p_1 states that p_2 is false and p_2 states that p_1 is false. Determine whether p_1 is true or false.

f. $2 \times 2 = 4$; therefore a given consequence does not hold. Determine whether or not Example f is true or false (under the assumption that Example f is unique).

g. We are given two premises:

g_1. "man is an animal"

and

g_2. any statement except g_1 is true.

Determine whether or not Premise g_2 is true.

h. We are given the following premises:

h_1. man is an animal,

h_2. the earth is round,

and

h_3. every statement except h_3 is true.

Is Premise h_3 true?

i. We are given the following premises:

i_1. twice two is four,

i_2. unicorns exist,

and

i_3. every statement other than i_1 and i_2 is false.

Determine whether Premise i_3 is true or false.

We will obtain paradoxes for Examples f through i, beginning with Example f. Assume that Example f is true; in this case the antecedent of Example f can be eliminated as a fixed true statement. But the assertion of the truth of the consequent of f is equivalent to asserting that all of Proposition f is false. As a result, f is false. We now assume that Example f is false, a condition that can occur only if the antecedent is true (and this is indeed so) and the consequent of f is false. But the assumption that the consequent is false is, in this case, equivalent to asserting that the entire proposition f is false. As a result, Example f is true. Paradox.

Examples g and i are of the same type in the sense that they include separative exponibilia (with the word "except"). They can be reduced to paradoxes in exactly the same way, so we will limit our analysis to Example h. Under the assumption that Premise h_3 is true, we can easily show that h_3 is false. If, however, we assume that Premise h_3 is false, that is, at least one statement in addition to h_3 is false, we can easily show that this is impossible since Premises h_1 and h_2 are both true, and, except for them and h_3, we have no other statements at our disposal. As a result, we must assert that Premise h_3 is true (reasoning from the general principle that, if an assumption leads to a contradiction, the given assumption is false).

Albert of Saxony presents a number of examples that, instead of containing insolubilia, contain only some system of premises composed of inconsistent statements. For example, we are given two premises *A* and *B*, where *B* denotes some fixed false statement and *A* states that "everything but *A* is different from *B*." It is required to determine whether *A* is true or false. Initially, assume that *A* is true, that is, the statement "only *A* does not differ from *B*" is true. We must also conclude that *B* differs from *A* (since the universe of discourse consists of only *A* and *B*), and this is impossible (it violates the law of identities). As a result, the assumption that *A* is true is untenable; thus, *A* is false. If we assume that *A* is false, we can only conclude that *B* is not different from *B*; that is, the statement "*A* is true," required to obtain a paradox, does not appear. Thus, the system of premises consisting of *A* and *B* is inconsistent, and analysis of it proves that *A* is false.

Buridan is responsible for a curious system of statements in which no statement is true, the system is not paradoxical, and each of them is defined in terms of another by means of the same method. Namely, he calls attention to the case of two premises, p and g, where p says: "g is false" and g says "p is false." It is easy to see that the assumptions

I. p is true and g is false and
II. p is false and g is true

are consistent. There is no way of deciding whether Assumption I is true or Assumption II is true.

Thus far, our discussion of antinomies has remained within the framework of two-valued formalism. The problem of how the Scholastics formalized modal antinomies (i.e., paradoxes whose statements contain modal terms) remains unanswered. Albert of Saxony, for example, presents an antinomy whose literal expression contains such expressions as "Socrates doubts that . . .," and "Plato proposes that . . .," (Reference 17, vol. 4, p. 80; Anmerkung, Reference 77). These can scarcely qualify as modal antinomies in the narrow sense since they can apparently be expressed completely within the framework of two-valued formalism. Albert himself offered the opinion that the nature of such antinomies is linguistic and psychological rather than logical.

Discussions of paradoxes in Scholastic logic were very lively. A classification of the viewpoints on the nature of insolubilia is contained in the work of Johann Mayoris Scott (1478–1540), who lived and taught in Paris (Reference 17, vol. 4, p. 250).

A close examination of the opinions (of which Scott counted eight and Bochenski (Reference 370) counted thirteen) shows that there were essentially three approaches: (a) "rejection" (*cassatio*), (b) "restriction" (*restrictio*), and (c) "resolution" (*solutio*).

According to Approach (a), an *insolubilium* is not a proposition (Paul of Venice) and, because it cannot be said to be true or false, it is simply meaningless. We will illustrate Approach (b) with an example taken from the work of Ockham. According to Ockham, the source of an antinomy lies in the fact that the terms required for notation of propositions are sometimes used for notation of the same propositions in which they are used as constituents. More plainly, Ockham meant that part of a proposition (the predicate "is false") must not (in order to eliminate an antinomy) refer to the entire assumption in which it appears.*

Ockham's view therefore reduces to an interdiction of circular definitions. In other words, it is not permissible to require linguistic constructions in which, for example, a given proposition appeals directly to falseness proper (or unprovability). By eliminating circular arguments in his studies of paradoxes, Ockham was led to his famous "razor" ("essentials should not be multiplied more than is necessary"). He believed that his solution to the problem was the most general possible (in the sense that if his proscription is followed, antinomies do not appear). However, Buridan found an example in which Ockham's proscription is not violated (there is no direct appeal to falseness), but an antinomy nonetheless appears.

* The problem of this type of restriction did not appear in connection with the predicate "is true," because, in defining truth, Ockham, following Aristotle, assumed that the propositions (1) "p" and (2) "'p' is true" are equivalent.

The system Σ consists of the following statements:

C_1. "man is an animal,"
C_2. "only C_1 is true,"
C_3. "C_1 and C_2 are the only available statements."

(See Example c in the discussion of the antinomies of Albert of Saxony; in that example, C_2 is an insolubilium, although it does not appeal directly to falseness.)

The system Σ therefore contains a paradoxical statement, C_2. The special property of C_2 is that it does not contain a direct appeal to the falseness of C_2. Rather, it refers only to Premise C_1. Without knowing anything about C_1, it would still be impossible to draw definite conclusions about the truth or falseness of C_2. Indeed, if C_1 stated that "the meaning of C_1 is identical to that of C_2," no antinomy would appear.

In view of this type of expression, Buridan distinguished two types of paradoxical statements: The first type contains "direct reference" (relative to the truth value ascribed to a proposition in accordance with the Aristotelian definition of truth), while the second type contains "indirect reference." For instance, in the liar's paradox, there is "direct reference," while in the above system of propositions, C_2 contains an indirect reference.

The existence of "indirect reference" provided Buridan with the occasion to declare Ockham's approach to elimination of semantic antinomies ineffective and to attempt to find other methods for elimination of paradoxes. In modern terms, Ockham's approach leaned toward Russell's theory of logical types, while Buridan's approach is somewhat reminiscent of Tarski's viewpoint in *The Concept of Truth in Formalized Languages* (Lvov, 1935).

Of the attempts made during the Middle Ages to eliminate antinomies, we will discuss in more detail only Buridan's theory since this approach is apparently the one of most value. It is presented primarily in his works *Sophismata* and *Johannes Buridani Quaestiones in Methaphysicam Aristotelis* (Paris, 1518).

From the formal viewpoint, Buridan's considerations on a method of eliminating semantic paradoxes in *Johannes Buridani Quaestiones in Methaphysicam Aristotelis* (VI, Qu. 11) can be stated as follows. We assume that p is some unsolvable proposition such as the one we considered above: "all that is written in this book is false." To eliminate the antinomy appearing in this case, Buridan proposes to append to p (the symbol for the given paradoxical proposition) an auxiliary premise Δp, where Δp states that the proposition p is actually spoken by someone, say Socrates. From the conjunction p & Δp, we can derive the proposition p. But from p, in view of the value of p, that is, the actual words in the proposition, we can derive the statement $\bar{p}$.

Thus, from p & Δp we can obtain the two propositions p and $\bar{p}$; that is, we can obtain an explicit formal contradiction. But, if a given statement leads to a contradiction, it is false. That is, the expression $\overline{p \,\&\, \Delta p}$ is true, but $\overline{p \,\&\, \Delta p}$ implies that $p \supset \overline{\Delta p}$. In other words, the proposition that p is true implies only that Socrates cannot be responsible for it. We should note that $\overline{p \,\&\, \Delta p}$ implies $\Delta p \supset \bar{p}$; that is, the premise that Socrates has spoken p implies that p is false. Thus, there is no paradox.

Buridan's ideas can also be formalized on the basis of an analysis of texts that are not in Prantl's compendium but are presented in Reference 20 (which gives several excerpts from Buridan's works). This latter reference initially attempts to find a concrete form of the Aristotelian meaning of the predicate "to be true" and presents the following method of formalizing Buridan's ideas.

Buridan criticized the viewpoint according to which all (including, therefore, paradoxical) propositions imply another proposition in which the subject is the name of the initial proposition itself and the predicate is "is true." If $\forall$ is the universal quantifier, x is some proposition, x_n is the name of this proposition, and T is the predicate "is true," then the proposition Buridan had in mind takes the form

$$\forall x(x \supset T(x_n)). \quad (1)$$

From the viewpoint of material implication, Formula 1 is, of course, irrefutable. While x may be false in Formula 1, the entire implication is nonetheless true. However, Buridan used the following argument to avoid the thesis that Expression 1 is always true. Information may actually be contained in x, but it need not be fixed in a particular statement. Hence x is only the name of a statement ("the consequent [that is, $T(x_n)$ in Formula 1], because it is affirmative, can denote nothing," said Buridan). He added an additional premise, asserting the existence of the proposition x. Thus, the expression "x_n is the name of x and x is" is equivalent to the phrase "x_n is true." Symbolically, we have

$$T(x_n) = x_n x \;\&\; x, \tag{2}$$

where the expression $x_n x$ is taken to be a single, inseparable symbol denoting the statement "x_n is the name of x" ("x_n denotes x"). Definition 2 can easily be used to eliminate the liar's paradox. Let some paradoxical statement (we call it m) mean (after it is appropriately interpreted) "m is false." Symbolically, we have $m = \overline{T(m_n)}$, where m_n is the name of m.

If we accept Thesis 1, a paradox is obtained as follows:

$$m = T(m_n) \qquad \text{(by 1)}, \tag{3}$$

$$m = \overline{T(m_n)} \qquad \text{(premise)} \tag{4}$$

It follows from (3) and (4) that $(Tm_n) = \overline{T(m_n)}$; that is, we have an antinomy that is easily eliminated by use of Premise 2. Assume that m is true. Then we obtain

$$T(m_n) = m_n m \;\&\; m. \tag{5}$$

By (4), we have that

$$T(m_n) = m_n\overline{T(m_n)} \;\&\; \overline{T(m_n)} \supset \overline{T(m_n)}$$

or

$$T(m_n) \supset \overline{T(m_n)}.$$

We now assume that m is false, i.e., that we have

$$\overline{T(m_n)} = \overline{m_n m \;\&\; m} \tag{6}$$

or

$$\overline{T(m_n)} = \overline{m_n T(m_n) \;\&\; (\overline{T(m_n)}} \tag{7}$$

from which we obtain

$$\overline{T(m_n)} \supset m_n \overline{T(m_n)} \vee \overline{\overline{T(m_n)}}. \tag{8}$$

It is now impossible to derive $T(m_n)$ from $\overline{T(m_n)}$, so a paradox does not appear.

Thus, the assumption that m is false leads only to the statement "either m is true or m_n is not the name of a true proposition m," but not to the categorical statement (required for an antinomy) "m is true."

In addition to Buridan's theory of elimination of paradoxes, we should also note a similar study by Paul of Venice; we will present a discussion of these ideas, somewhat modifying the formalization presented in Reference 370, pp. 291–292. Paul assumed that, because an insolubilium is a proposition, there is no basis for assuming that the Aristotelian criteria of truth do not extend to it. To eliminate antinomies, however, it is necessary to verify the difference between the "ordinary" and "precise" value of a paradoxical proposition. We will see that this requirement by Paul is equivalent to a refinement of the Aristotelian notion of truth and, in connection with this, that Paul's ideas did not extend past the basic superstructure constructed by Buridan.

Before proceeding to what Paul actually meant, we introduce the following conventions. By f we denote the falseness functor, and by t we denote the truth functor. We will write an insolubilium in the form

$$A \equiv (A \equiv f). \tag{1}$$

Adding Paul's refinement of Aristotle's notion of truth, we obtain the following system of premises:

$$(A \equiv p) \rightarrow ((A \equiv t) \equiv ((A \equiv t) \,\&\, p)), \tag{2}$$

$$(A \equiv p) \rightarrow ((A \equiv f) \equiv \overline{((A \equiv t) \,\&\, p)}), \tag{3}$$

to which we must add the usual definition of falseness,

$$(A \equiv f) \equiv \overline{(A \equiv t)}. \tag{4}$$

We leave it to the interested reader to verify that Premises 1 through 4 and the assumption that A is true imply that A is false. On the other hand, let us consider the consequences of Premises 1 through 4 under the assumption that A is false. First of all, substituting $A \equiv f$ for p in Premise 3, we obtain

$$(A \equiv (A \equiv f)) \rightarrow ((A \equiv f) \equiv \overline{((A \equiv t) \,\&\, (A \equiv f))}). \quad (5)$$

Application of *modus ponens* to 1 and 5 yields

$$(A \equiv f) \equiv \overline{(A \equiv t) \,\&\, (A \equiv f)}. \quad (6)$$

Since we have assumed that A is false, 6 reduces to

$$\overline{(A \equiv t) \,\&\, (A \equiv f)}, \quad (7)$$

which is equivalent to the expression

$$\overline{A \equiv t} \vee \overline{A \equiv f}. \quad (8)$$

But, by Premise 4, $\overline{A \equiv t}$ is equivalent to $A \equiv f$. Therefore, finally we obtain

$$(A \equiv f) \vee (\overline{A \equiv f}), \quad (9)$$

that is, a variation of the principle of *tertium non datur*, from which naturally we cannot derive the necessary conclusion (for a paradox) that A is true. Thus, the paradox has been eliminated, and, according to Paul, the assumption that the insolubilium is false leads only to a statement about the validity of the law of the excluded middle.

It should be noted that many of the viewpoints of medieval scholars on the nature of insolubilia anticipate in one way or another some of the contemporary approaches to elimination of antinomies. For example, some Scholastics insisted on strict prohibitions against the "vicious circle" (*circulus vitiosus*), which they saw as an immediate source of paradoxes. In the opinion of others, an "insolubilium is neither true nor false but lies somewhere in between these notions, different from both,"* an opinion, in modern terms, equivalent to the statement that paradoxes must be resolved in trivalent logic.

* For the text of Paul of Venice on the opinion of his opponents on the subject of insolubilia, see Reference 370, p. 281.

By the end of the sixteenth century, the emphasis on logical research had diminished. And, about the same time, it appears that the differentiation between the logical schools became complete. At the end of the Middle Ages, there were basically three logical schools: (1) the school of peripatetics (which had its origins in the *Summulae* of Peter of Spain); (2) the Ramists, the followers of Peter Ramus (1515–1572), who persistently advanced proposals for reforming Aristotelian logic; and (3) the Lullists, who based their doctrine on the studies of Raymond Lully.

The Frenchman Peter Ramus was not satisfied with the level of Aristotelian logic. Following Lorenzo della Valle and other Stoics, he sharply censured the gap between logic and rhetoric. A number of like-minded scholars in Germany, for example, Johannes Sturm (1507–1589), were drawn to Ramus' criticism.

Ramus criticized the Aristotelian system of categories, which, in his opinion, did not permit classification of logical methods. He had some influence on the logic of Leibniz (in particular, on the problem of using certain figures of syllogisms for proving particular rules concerning the conversion of sentences). Mainly, Ramus proposed using syllogisms of the second and third figures for this purpose. For instance, the modus *Datisi* of the third figure and the principle of identity can be used to prove the reversal rule that SiP implies PiS (that is, the statement "some S are P" implies that "some P are S"). Indeed, if we substitute the term P for M in the statement $(MaP.\ MiS) \rightarrow (SiP)$, we obtain $(PaP.\ PiS) \rightarrow (SiP)$, from which, by appealing to the identically true statement PaP, we obtain $PiS \rightarrow SiP$.

An important place in the logic of Ramus was occupied by problems of methodology and, in particular, the theory of loci, (*loci communes*), that is, points of view that make it possible to construct proofs. He distinguished five "original" points of view and nine "derived" points of view.

Of the works of Ramus, we should note the following: *Animadversiones in dialecticam Aristotelis* (1543); *Dialecticae partitiones* (1543); *Institutiones dialecticae* (Paris, 1543, 1547); *Scholae in liberales artes* (Basel, 1569, 1578, 1582).

The logical ideas of Lully were completely adopted by the well-known Spanish pedagogue and humanist Luis Vives (1492–1540) and the great thinker Giordano Bruno (1548–1600).

Vives insisted that scientific research be based not on the authority of Aristotle but on experiment. In his works *On the Means of the Probable* and *Against Pseudodialectics* (1519) he argued against exaggerating the value of the rhetorical applications of logic, preparing the way for the empirical methodology of Bacon. Vives was responsible also for an ideographic representation of the relations between syllogistic notions. The diagram in Figure 7 is

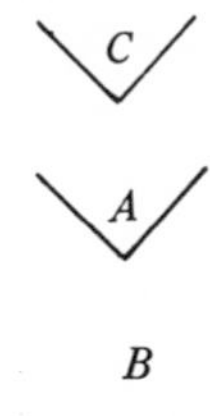

Figure 7

reproduced from his *De censura veri* (1555), where *C* denotes the minor premise of a syllogism, *A* denotes the middle, and *B* is the major term; the inclusion relation is indicated by the V-shaped sign.

Vives' work can be treated as a precursor of ideographic methods later developed by Hamilton. In 1555 his dialogues were translated into German by Breringen in Oldenburg. The collected works of Vives were published that same year in Basel and in 1782–1790 in an eight-volume edition in Valencia. In the Russian language we have *A Guide to Knowledge* (St. Petersburg, 1768), which was translated from the Latin. Further information about Vives can be found in: A. Lange, *Ludvig Vives* and *Schmidt'schen Encyclopädie des Erziehungs- und Unterrichtswesens*, vol. 9 (1869).

A number of ideas in the propositional calculus were slowly formulated by the Lullists in the special study called "axiomatica" that was used as an introduction to syllogistics. In particular, this is the way J. H. Alsted proceeded in his handbook *A New Form of General Logic* (1652). Because he strictly adhered to certain logical ideas of Ramus, Alsted attempted to reconcile the dialectics of

Ramus and Lully with Aristotelian logic. Born in 1588 in Nassau, he was appointed professor of philosophy at Strassburg in 1610. Alsted had commented on Lully's work in *Sources of Lully's Art and True Logic* in 1609. In 1624 he made an attempt at further development of the ideographic methods of Vives. In one of his Frankfurt works, written about 1641, Alsted attempted to state the axioms basic to a number of the philosophical systems of his contemporaries.

Bruno was an active commentator on Lully, as we can see from his works *De compendioso architectura et complemento artis Raymundi Lullii* (Paris, 1582); *De Lulliano specierum scrutinio* (Prague, 1588); and *De Lampade combinatoria Lulliana.*

In keeping with his fondness for convenient mnemonic devices, Bruno attempted to justify the logical technique of the *Ars Magna* by gnosiological apparatus. In *De Lampade combinatoria Lulliana* he considered three forms of activities of the human mind: perception of elementary objects, identification and distinction of objects, and argumentation. Accordingly, in *Ars Magna* a theory for a method of combining terms and argumentation based on the laws of combining terms is derived from a list of initial, undefined terms in a given alphabet.

Bruno was also responsible for a detailed critique of the Scholastic Peripatetic school of logic from the viewpoint of his somewhat improved version of Lully's principles. In his *On the Shadows of Ideas* (1582), he raised the more general problem of improving memory.

A final important achievement of the Lullists was their simplification of the cumbersome studies of Peter of Spain on the forms of suppositions.

The foregoing discussion, of course, raises the puzzling problem of the extent to which the considerable logical achievements of the Middle Ages were forgotten during the Renaissance. Before we attempt to answer this question, we should note the fact that there was an objective limit to further evolution of the Scholastic form of logic. Just as Greek geometry could not advance beyond Apollonius since literal algebra had not yet been developed, so the Scholastic logic of the Middle Ages could not advance beyond its state in the *Logica Magna* of Paul of Venice since no formalism could be found that would make possible consolidation of prior achievements, thus making possible further progress.*

Another source of the well-known *cul de sac* of Scholastic logic can be seen in the fact that the mathematics and technology of the Middle Ages could not sufficiently stimulate its development, not even to the point of developing effective symbolism for the formalisms. The decisive factor, however, is that the technology of the Middle Ages could not stimulate logic the way the industrial practice of rising capitalism could. By virtue of these circumstances, the individual great logical achievements of the Middle Ages, which were not based on an artificial "language," were sometimes formalized by extremely unreliable mnemonic devices and they thus necessarily remained deeply hidden.

Mathematical logic began to flower only in the nineteenth century under the influence of the exact sciences, which made complex computational problems the order of the day. Even the form of the new logic differed from Scholastic logic. Syntactical problems of logic were relegated to the background of semantic problems, which, for a time, were of absolutely no interest to scholars. For example, the Scholastic theory of suppositions was treated as being of doubtful logical value. To some extent, there has been a return to the syntactical form of ancient logic, although the modern form is no trivial reproduction since natural science and technology have advanced greatly, with consequent major influences on the evolution of logical thought. By no means can we assume that the logic of the Middle Ages has been completely analyzed. There are many "gaps," such as the problem of determining the major medieval conceptions of semantic implication and what modern forms of implication can be used to formalize them. In addition, we do not know whether it was only semantic

* It is quite reasonable to argue that the source of this circumstance is the underdeveloped technology and the social structure of Greece at the time of Apollonius (and later) or of the European Middle Ages. Certainly, industrial technology and the nature of the social structure are related to the development of science, which led to both algebra and formalisms not available to the Middle Ages. However, the immediate reason that Greek geometers and the logicians of the Middle Ages were unable to progress was their lack of the necessary research techniques.

paradoxes (and not other types of antinomies) that were analyzed. It can also be asked whether inductive logic developed during the Middle Ages.* This list can be greatly extended; it is only meant to indicate the necessity of further research on the logical heritage of the Middle Ages.

It is beyond the scope of this book to consider the immense subject of the fate of our logical inheritance from the Middle Ages. However, we should note, for example, that the problem of the extent to which Western European logic affected the development of logical thought in Russia and in other Eastern European countries has received little attention. The Russian scholars M. V. Bezobrazova (Reference 14) and N. A. Sokolov (Reference 15) have discovered Russian manuscripts of the seventeenth and eighteenth centuries that contain extensive critical commentaries on Lully's *Ars Magna*. One variant of this manuscript was found by Sokolov in the Kazan library of the Solovetsky monastery.

It is interesting to note that initially the manuscript belonged to a peasant, Semen Ivanov, who lived on the holdings of the Solovetsky monastery close to Archangel; it was sold to a servant, Grigory Titov of the Solovetsky monastery (Reference 15, p. 332). Thus, it would seem that there were many copies of the Russian interpretation of Lully's *Ars Magna*. One of the copyists has called it a "fragrantly productive garden."

It was demonstrated in References 14 and 15 that the author of the Russian interpretation of *Ars Magna* was a translator of the "ambassadorial office," Andrei Khristoforovich Belobodsky (Reference 15, p. 337). According to Savitskiy (*Trudy Kievskoy Dukhovnoy Akademii*, November 1902, p. 444), Belobodsky "discovered the logical principles of Lully completely independently." As Trakhtenberg has properly remarked, Belobodsky stated as his aim an attempt to find a universal method for obtaining knowledge, using graphical methods for this purpose (a table is a "grid"

* Of course, individual remarks on inductive logic can be observed in the works of the scholars of the Middle Ages. For example, Lully considered induction (*inductio incompleta*), and Duns Scotus asserted that a defect in Aristotelian induction existed in Aristotle's thesis that, in the process of perception, the individual precedes the general.

classifying "all reality," the "chain of being," etc.). See O. V. Trakhtenberg, "Social and Political Thought in Russia During the fifteenth through seventeenth centuries," in *On the History of Russian Philosophy* (Moscow, 1951), p. 87.

The logical ideas of Lully were being introduced into the Secondary schools of Russia at that time, and they penetrated even into rhetoric. However, the influence of these ideas was not strong and waned rapidly. The same must be said of the influence of Scholastic methodology on Russian academic philosophy in the seventeenth century.

CHAPTER TWO

Leibniz, The Founder of Symbolic Logic

1. The Forerunners of Leibniz

The center of attention in scientific thought during the seventeenth century was focused on the creation of general methods for solving the ever-widening classes of problems posed by the rapid advance of exact science. One such method to appear was algebra, which began the qualitative extension of the means available for scientific (and later, logical) research.

As early as 1623, Wilhelm Schickard, professor of astronomy and mathematics and a friend of Kepler, built the first computing machine. This discovery actually dealt a powerful blow to the religious–idealistic notion of thought since it demonstrated that human thought could be stimulated by an automatic device (at least in some respects).

Among the thinkers of the seventeenth century who addressed themselves to the problems of mathematical logic prior to Leibniz' time, we should note the German Joachim Jung and the Belgian philosopher Arnold Geulincx. While we can say nothing definite about the influence of the latter on Leibniz, it is certain that Leibniz was directly and fruitfully influenced by the former.

Joachim Jung was born into an academic family on October 21, 1587 in Lubeck and died on September 23, 1657 in Hamburg. At Rostock he studied philosophy, mathematics, and natural science, earning his master of arts degree in 1609. At twenty-two he became professor of mathematics at Rostock, and in 1625 he was appointed professor of medicine at Helmstadt. In 1626 he was again appointed professor of mathematics at Rostock, and in the same year he moved to Hamburg to become the rector of the local academic secondary school.

Jung's scientific interests were concentrated primarily in the fields of mathematics and physics. According to Leibniz, Jung was responsible for demonstrating the incorrectness of Galileo's assertion that the figure described by a string freely suspended by its ends is a parabola. He was actively concerned with introducing mathematical methods into natural science as well as with demonstrating the necessity of quantitative measurements of empirical processes.

Of the works of Jung we should note the following: *Logica Hamburgensis* (Hamburg, 1638) and *Doscoscopial physical minores* (1662). In the latter treatise, the author shows that he is a proponent of the empirical methodology of Francis Bacon, even though he leans toward a mathematical treatment of scientific and theoretical problems. Jung attempted to overcome Bacon's underestimation of the value of mathematical methods in research.

The problem of creating a logical system oriented toward scientific applications was proposed in its general form by Descartes and Pascal and was concretely developed first by Jung in his *Logica Hamburgensis* and later by Leibniz.

Before proceeding, we will briefly summarize the logical results of Blaise Pascal (1623–1662), that great mathematician and proponent of Cartesian methodology who can be considered as one of the predecessors of the modern axiomatic method. He was responsible for refining a series of problems in the logical theories of definition and proof. According to Pascal (see *On a Geometrical Language*), any proof must satisfy the following three requirements: (1) the terms that are used must be clearly defined; (2) premises or obvious axioms must be used; (3) in the course of a proof, defining terms must be substituted for terms being defined.

The research of Pascal in logic provided the basis for studies of the logicians of Port Royale. Pascal was known for his witty critique of the probabilism of the Jesuits, a critique he wrote from the Jansenists' position. Alonzo Church cites Pascal's work among the predecessors of the so-called optative logic (literally, the logic of indebtedness), a construction that is absolutely consistent with modern mathematical logic. Literature on Pascal as a logician and a mathematical methodologist may be found in References 83–89.

In 1638 Jung focused his attention on the necessity of constructing a logical system similar to the mathematical calculus. He demonstrated that classical syllogistics did not cover certain conclusions actually encountered in science, and he attempted to construct a theory of inherently nonsyllogistic conclusions, a theory later characterized as part of the theory of relations. As an elementary type of conclusion that cannot be formalized by syllogistic means, Jung presented the following example: If A is the father of B, then B is the son of A. Here are some other examples of nonsyllogistic sentences considered by Jung: "Conversion of relations. David is the father of Solomon and Solomon is the son of David." "A circle is a figure. Therefore, whenever a circle is drawn, so is a figure."

The best testimonial to Jung is presented by the following statement of Leibniz: "Among all who have ever undertaken an attempt at developing a true art of proof, I know none other to have penetrated the subject more deeply than Joachim Jung of Lubeck" (cited in Reference 92, p. 146).

An important place among the mathematicians responsible for the development of symbolic logic is occupied by the Belgian logician and philosopher Arnold Geulincx. Born in Antwerp in 1625, he studied theology and philosophy at Leuven, where he began to teach in 1646. As the result of his criticism of Scholastic philosophy, he became the object of sharp attacks by the Church. Clerical denunciation caused him to be dismissed from his post, and for a long time he retired to private life in Leyden. However, in 1665, Abraham Heyden found him a post as professor at the local university, where he remained until his death in 1669.

In the theory of knowledge, Geulincx followed Descartes. Those of his important works that had a direct relationship to his philosophical views were published only after his death. We note the following: *Saturnalia sive quaestiones quodlibeticae* (1660); *Logica fundamentis suis restituta* (1662); the incomplete *Γνωϑι σευτὸν sive Ethica* (1665); *Compendium phisicum* (1688); *Annotata praecurrenti in Cartesium de principiis philosophiae* (1690); and *Metaphysica vera et ad mentem peripatericam* (1691).

In his treatment of the fundamental philosophical problem of the relationship between the soul and the body, Geulincx abandoned the Cartesian viewpoint and stated a hypothesis that has become known in the history of philosophy as the theory of synchronism (parallelism) in the behavior of the soul and the body. According to Descartes, man is a machine with a "soul," which, coming into contact with the "vital spirits," affects the operation of the body. This is the approach that Geulincx abandoned. He began with the physicists' discovery of the law of conservation of momentum, which states that not only is the total momentum in the universe constant but so is the total momentum in any given direction. Physical action is associated with impact. According to Geulincx, the laws of dynamics are completely sufficient for description of any motion of matter. Thus, Descartes notwithstanding, it is impossible for the mind to have any effect on matter.

This raised the problem of how to reconcile the fact that the human body behaves as if it were controlled by the mind. The answer provided by Geulincx to this problem is paradoxical: Consider a pair of synchronously operating clocks. If we look at one clock and listen to the other, we automatically assume that the first causes the second to operate. In fact, however, there is no causal relationship between the two clocks, although they both always show exactly the same time. Precisely the same relationship holds between material and psychological phenomena — they are totally independent. One substance (the soul) cannot affect the other (the body). However, according to Geulincx, a statement about an ideal phenomenon can be rigorously and uniquely predicted by means of the equivalent statement about the corresponding material event.

Although Geulincx' theory is far from the subsequent materialism, it undoubtedly would be unacceptable to theologians, especially the last assertion.

In 1662 Geulincx published his work *Logica Fundamentis suis, a quibus hactenus collapsa fuerat restituta*, in which several chapters deal with mathematical logic. The basic scheme for the logical proofs is the same as that upon which the proofs given in Euclid's *Elements* are based.

What sources other than Euclid did Geulincx have in his attempts to reform Scholastic logic? We have no documentry evidence that indicates an influence of Jung on Geulincx, and it is possible that the latter did not even know of his German predecessor. On the other hand, it is unlikely that Geulincx was not familiar with Hobbes' (1588–1579) *Leviathan*, which contains a general form of the ideas of logical calculus (pt. 1, chap. V). *Leviathan* appeared eleven years prior to the publication of *Logica Fundamentis suis* in 1651. According to Hobbes, thought is nothing but a calculus in which words play the value of conventional symbols or counters; it is an error to think that counters are used only for money. Hobbes also stated that mathematics must be treated as a model for any type of knowledge of the world. We must not think, he said, that calculations in the finer sense deal only with numbers; we add and subtract not only magnitudes but objects, motions, time, qualities, acts, relations, propositions, and words (in which any kind of philosophy is contained) (see Hobbes' *Collected Works*, Moscow–Leningrad, 1926, pp. 7–8). Elsewhere, Hobbes said that just as we add and subtract numbers in arithmetic, in logic we do the same with words: adding two words, we obtain a statement; adding two statements, we obtain a syllogism. And several syllogisms form a proof (*Leviathan*, pt. 1, chap. V). Thus, through his methodological arguments, the materialist Hobbes opened the way for later logical and mathematical conceptions.

Geulincx stated that it is possible to construct an infinite set of statements, and one of the infinite subsets of this set is the class of true statements. He stated and proved a number of theorems concerning the propositional calculus. Among them we should note the rules that later became known as De Morgan's laws and a theorem asserting that if a conjunctive relationship is proved between two statements, a disjunctive relationship is also proved. Symbolically, we have $(a \,\&\, b) \rightarrow (a \vee b)$. The following theorems of Geulincx are also interesting:

1. $\overline{A \rightarrow B}$ implies $\overline{\bar{B} \rightarrow \bar{A}}$.
2. $B \rightarrow C$ and $\overline{A \rightarrow C}$ implies $\overline{A \rightarrow B}$.
3. $A \rightarrow B$ and $\overline{A \rightarrow C}$ imply $\overline{B \rightarrow C}$.

4. A & B implies $\overline{A \rightarrow B}$, where the bar denotes logical negation.

Geulincx rejected the syllogism

Every white man is white.
Every white man is a man.
―――――――――――――――
Consequently, some men are white.

on the basis that the premises are necessary while the conclusion is contingent. After all, a contingent statement cannot follow from necessary statements. It is curious that here Geulincx has rejected the modus Darapti, a syllogism recognized as not generally meaningful only several centuries later (although this later rejection was made for other reasons).

2. A Brief Biography and Survey of the Works of Leibniz

The founder of symbolic logic, Gottfried Wilhelm Leibniz, was born on June 21, 1646 in Leipzig. His father was a professor of moral philosophy and held the position of Registrar. Young Gottfried made full use of his father's large library. At twelve, he had mastered Latin and began the study of Greek.

At fourteen, Leibniz wrote:

"I saw that in logic simple notions are separated into definite classes, namely, so-called predicaments, and I was astonished that composite statements or sentences were not divided into classes according to a system in which every term can be derived from another: these classes I called the predicaments of sentences, which then provide the material for conclusions, just as ordinary predicaments provide the material for sentences. . . . Later, I saw that the system I required was the same as that used by mathematicians in elementary studies, where they place their statements in an order such that each statement follows from the previous ones. It was precisely this that I then vainly attempted to find in philosophy" (cited in Reference 96, p. 35).

Leibniz continued thus:

"Applying myself with great zeal toward these ends, I necessarily came upon the surprising idea that it might be possible to find some alphabet of human thought and that, by combining the letters in this alphabet and analyzing the words thus composed, it might be possible to derive and discuss everything. When this thought was born within me, I was quite excited, but, of course, this was an infantile joy; at that time I still could not comprehend the tremendous scope of the matter. However, the more I learned, the more difficult it became for me to provide a solution for so large a problem." (Cited in Reference 96, p. 39).

Here we have an idea quite similar to the illustrious principles presented by the Spanish Scholastic Raymond Lully (1235–1315) in his *Ars Magna*. Leibniz later proceeded far afield of this initial principle and developed an idea incomparably more significant than the comparatively simple conceptions of Lully.

At fifteen Leibniz began to study at the university of Leipzig, devoting his attention primarily to the study of law. At the same time, he studied philosophy (under Jacob Tomasius) and then mathematics in Geneva (under the physicist Erhard Weigel (1625–1699), who familiarized his student with Euclidean geometry and elementary algebra. In 1664 Leibniz won the degree of Master of Arts and in 1666 obtained his doctorate in law at Altdorf. In 1667 he entered the service of the Elector of Mainz. Initially, he wrote commentaries on juridical themes concerned either with so-called "problematical cases" or problems that were hidden beneath the unique logic of fine juridical definitions. In dealing with difficult cases of jurisprudence, Leibniz defended the thesis that it is necessary to supplement formal laws by meaningful appeals to common sense.

On the occasion of refusing a proffered extraordinary professorship, Leibniz became acquainted in Nuremberg with a former Minister of Mainz, Baron von Boyneburg. After a short period in Frankfurt-on-Main, he returned to Mainz and the court of the elector, where he took part in the practical work of reforming legislation. In Paris, Leibniz became acquainted with the French materialist Pierre Gassendi and the Dutch physicist Huygens. It was here that he began his fruitful work in mathematics. An acquaintance with the mathematical research of Pascal and Huygens greatly enriched his knowledge.

In 1672 Leibniz improved Pascal's computer, after which,

apparently in 1673, he was made a member of the Royal Society. At this time he became intensely interested in problems of dynamics. He stated the law of conservation of force, which he used as a basis for studying the laws of dynamics.

In 1673 Leibniz developed his general method of tangents (in connection with the general problem of finding the tangent to a curve). Shortly after this (in 1675–1676) he developed (independently of Newton) the fundamental principles of the differential calculus (first published in *Acta Eruditorum*, 1684). In 1679 he extended to logic the ideas of Viète and Descartes concerning literal algebraic calculus.

In *On the Secrets of Geometry and Analysis of Indivisible and Infinite Quantities* (1686), Leibniz first used the integral sign (the word integral itself was introduced by Bernoulli). In a letter to L'hôpital in 1693, Leibniz used multiple indices to state the result of three linear equations. He also worked on the problem of elimination (in the general theory of equations) and laid the foundations of the theory of determinants. In a letter to Wallis (1698), Leibniz suggested expanding cube roots into infinite series.

In 1702 Leibniz published *A Justification of the Calculus of the Infinitely Small*, in which he attempted to justify his algorithms for differentiation and integration. In 1702–1703 he integrated rational fractions in trigonometric and logarithmic functions. In the theory of special curves Leibniz proved, in *Acta Eruditorum* (1706), that the evolvent is obtained by suspending a curve along the evolute. Somewhat earlier, he found the characteristic property of the tractrix and, independently of Gregory, he stated the equation of the spherical loxodrome, which is important in navigation. Finally, Leibniz introduced the term "function," which Bernoulli borrowed from him.

Leibniz actively published both mathematical and philosophical works in the Paris scientific journal *Journal de Sçavants* and the Leipzig *Acta Eruditorum*. He died November 14, 1716.

In his *Dissertation on the Combinatorial Art* (Leipzig, 1666) Leibniz first discussed his conception of the "universal characteristic," one aspect of which was to develop the ideas of pasigraphy as an attempt at creating a universal language of science. This project was begun in a period when Leibniz had not yet creatively contributed to mathematics.

A central idea of Leibniz' combinatorial system is the relationship between the "whole" and its "parts." Another important notion is expressed by the term "position" ("disposition"). The symbolism used in this work is also of interest. For example, instead of the expression "a combination of three elements," Leibniz wrote *com* 3 *nationem.*

Finally, Leibniz gave a formal definition of a series of extralogical relations, such as "is the father of," "is the son of," and "is the brother of." For example, from "a brother is the son of the father," *father* and *son* can be combined to form the notion of *being a brother.*

In his *Reflections on Knowledge, Truth and Ideas* (1684), Leibniz presented a classification of concepts (ideas) from the viewpoint of their certainty (hazy and clear, vague and distinct, symbolic and intuitive, etc.), and he developed a theory of analysis (in the formal logical sense of this word) according to which the proof of a proposition is equivalent to demonstrating that the subject and predicate are identical. He called proof the process of operating with definitions by substituting equal things for equal things.

In his *Difficultates quaedam logicae* (*Oeuvres philosophiques latines et françaises de M. de Leibniz, tirées de ses manuscripts qui se conservent dans la Bibliothèque royale de Hanovre*, Raspe, 1765), he presented a logical theory in which the predicate of any statement is always only part of the subject. He stated relations that express the relationship between universal and existential propositions.

1–8. *Nouveaux essais sur l'entendement humain par l'auteur de l'harmonie préetablie* (Raspe, 1765). Here Leibniz presented a critical discussion of the ideas set forth by Locke in his *Essays.* In particular, the critic demonstrated that Locke's empirical theory of abstraction is unsatisfactory, and he analyzed the basis of school logic. In parallel with his critique, Leibniz broadly developed his own philosophical views ("the theory of innate ideas," criteria for truth, etc.). These essays constitute Leibniz' fundamental philosophical work.

9. *Historia et commendatio linquae characteristicae universalis, quae*

simul sit ars inveniendi et judicandi (*Oeuvres*, Raspe, 1765, pp. 533 ff.). Here Leibniz developed one of his favorite ideas: "an alphabet of human thought that makes it possible to deductively derive new ideas by means of definite rules for combining symbols." Here the logical idea of pasigraphy is clearly distinguished from the linguistic idea of creating a "universal language."

10. *Unprejudiced Reflections on the Applications and Improvement of the German Language* (German manuscript prepared about 1697).

11. "Addenda ad specimen calculi universalis" (Reference 166, vol. 1, pp. 98–99).

12. "Non inelegans specimen demonstrandi in abstractis" (Reference 172, vol. 7 (1890), pp. 228–235). An English translation of this work appears in Reference 364, pp. 373–379. In particular, here Leibniz stated the logical law of idempotence: "*Si idem secum ipso sumatur nihil constituitor novum, seu* $A + A = A$" (if two identical symbols are added, nothing new appears, that is, $A + A = A$). Leibniz also considered the problem of expressing negative propositions, although he limited his interest primarily to a representation of the universal negative proposition.

13. "Fundamenta calculi ratiocinatoris" (Reference 166, vol. 1, pp. 92–94). Here Leibniz used the term "*calculus ratiocinator*" as an equivalent to the term "*logistica.*"

14. "Definitiones logicae" (Reference 166, vol. 1, pp. 100–101). According to Leibniz, a real definition must include all conditions necessary and sufficient for proving all properties of a given object. Leibniz rejected Hobbes' thesis that any definition must be nominal.

15. "De scientia universalis seu calculo philosophico" (Reference 166, vol. 1, pp. 82–85).

16. "Ad specimen calculi universalis addenda" (Reference 172, vol. 7, 1890, pp. 221–227).

17. *Elementa characteristicae universalis* (April 1679). Here Leibniz made his first attempt to construct a logical calculus. He presented an arithmetic interpretation of Aristotelian syllogistics, which he developed in the next work.

18. *Elementa calculi* (April 1679).

19. "Calculus consequentiarum" (Reference 165, pp. 84–89).

20. "Conversio logica" (Reference 165, pp. 253–255). Here

Leibniz investigated the difficulties associated with the operation of conversion for the purpose of expressing general propositions in existential form (when, for example, the proposition "any x is y" is taken to be equivalent to the expression "those x that are not y do not exist").

21. *Fundamenta calculi logici* (2 August, 1690). In contrast to the papers cited in Paragraphs 13 and 14, here Leibniz attempted to construct a logical calculus in algebraic rather than arithmetic form. He continued to develop his ideas in the next work.

22. "Logical Fragment" (Reference 172, vol. 7, 1890, pp. 236–247). An English translation of this text may be found in Reference 364, pp. 379–387.

23. "De formae logicae comprobatione per linearum dictus" (Reference 165, pp. 292–321). Leibniz presented a topological interpretation of the classical propositions of syllogistics in which he associated a line segment with each term. In addition, he developed the abstract theory of the notion of "inclusion."

24. "Specimen calculi universalis" (Reference 172, vol. 7, 1890, pp. 218–221).

25. *Generales inquisitiones de analysi notionum et veritatum* (1686). Here, in particular, Leibniz translated syllogistic propositions into existential form. This work proceeds further along the path of algebraic construction of a logical calculus.

26. "De ortu, progressu et natura Algebrae" (Reference 167, pp. 203 ff.). According to Leibniz, algebra must be subject to the universal characteristic, as a system of mathematical categories. A quantitative method is a method based on a system of mathematical relations.

27. *Monadologia* (1714). According to Leibniz, "our considerations spring from two great sources: (1) contradiction, by virtue of which we reason that what is false contains in itself a contradiction and that what is true opposes or contradicts what is false . . . and (2) sufficient reason, by virtue of which we say that no phenomenon can be said to be true or false, no assertion can be held to be valid without sufficient reason for us to say this is the way the matter stands, and not some other way, although in the majority of cases these reasons may be quite unknown to us." In addition, the

Monadologia contains the famous Leibnizian classification of true statements into truths of understanding and truths of fact.

28. "Methods of Universal Synthesis and Analysis" (G. W. Leibniz, *Philosophische Werke*, vols. 1, 2, Leipzig, 1924). Here Leibniz not only formally derived a given relation from axioms but also (semantically) interpreted it in terms of elements of his universal characteristic (*characteristica generalis*). The relation under discussion, which corresponds to a modern expression of the form $F(X, Y, Z)$, was not only formally derived from a given system of axioms Σ, but was also verified on elements $A_1, A_2, A_3, \ldots$ of the universal characteristic.

Further literature on Leibniz, as well as a list of the basic editions of his works, may be found in References 96–176.

The British historian Leslie Ellis was the first to call attention to Leibniz' role as a founder of symbolic logic.

3. The Methodological Principles of Leibniz

There were three major sources for the methodological views of the founder of symbolic logic: Lully's ideas about mechanization of human thought, Giordano Bruno's epistemology, and the ideas of Descartes about constructing a universal method for solving scientific problems, especially mathematical problems (the idea of a "universal mathematics").

Raymond Lully, the monk of Aragon, thought that it was possible to prove the dogmas of Christianity by purely logical means. He is responsible for not only the idea but for the first practical machine for mechanizing the process of logical derivation. Lully's machine took the form of a system of concentric rings on which notations for ideas were written. As the disks were rotated, the ideas were combined in a predetermined manner, which yielded syllogistic types of conclusions from given premises.

Lully proposed to construct all possible propositions from combinations of a given set of initial ideas. The number of such combinations is *a priori* $511^6 = 17{,}804{,}320{,}388{,}674{,}561$ (!) (this number also includes impossible combinations of terms). Because concentric disks were used, this number was actually about $9^6 =$

531,441 possible combinations of initial terms. Of course, Lully's machine was extremely limited, although this did not prevent the inventor from naïvely believing it could construct any conclusion and could find all true statements.

Not only did Lully's contemporaries not realize how revolutionary his ideas were (and, in fact, the importance of his studies became obvious only in the light of the scientific successes of the twentieth century), but they even mocked them. Only a comparatively small circle of Lully's followers continued to work on further developments of their teacher's ideas.

In his youth, Leibniz studied Lully's art (Reference 99, pp. 14 ff.). Thoughts similar to his combinatorial characteristic may also be found in a number of cabbalistic works. These works could possibly contain interesting logical ideas, just as one might find elements of the latest completely scientific notions of chemistry in various alchemical ideas. However, Lully's logical work had the greatest influence on Leibniz (see Reference 99, p. 16), and they led Leibniz to the development of a theoretical and applied pasigraphy.

Two aspects of Leibniz' "universal characteristic" should be distinguished. The first is associated with the linguistic idea of creating a "universal language," suggested to Leibniz by the British philologist George Dalgarno (1627–1688), who attempted in his *Ars signorum vulgo character universalis et lingua philosophica* (London, 1661) to tabulate ideas so that it would be possible to proceed from general classes to special subdivisions.

Another aspect of the "universal characteristic" consists in the requirement of developing a universal language and symbolism. According to Leibniz, we use signs not only to communicate our thoughts to other men but also to facilitate the very process of thought. Thus, he did not confuse the linguistic and logical aspects of his "universal characteristic." In symbolic form, operations on symbols should reflect all permissible combinations of the objects they denote, and they should make impossible combinations apparent.

Logical pasigraphy must follow the example of mathematics. Leibniz recognized the difficulties associated with realizing his ideas: (1) the concepts of the empirical sciences are not all so clear and obvious as certain mathematical notions; (2) the problem of finding the fundamental notions in certain sciences is extremely complex. Leibniz, however, was prepared to work as hard as necessary to create a universal logical method that would make it possible to replace argument by formal computation. Instead of involving themselves in unproductive disputes, philosophers would be able to take their pencils in hand and say, "Let us calculate."

Thus, the "universal characteristic," taken as a logical plan, is a system of rigorously defined symbols that can be used in logic and other deductive sciences to denote simple elements of the objects under investigation by a given science. First of all, these symbols would have to be brief and compressed in form, and they would have to include a maximum of information in a minimum of space; second, there would have to be an isomorphic correspondence to the objects they denote, so as to make it possible to represent simple ideas in as natural a manner as possible. Complex ideas would have to be representable as combinations of elementary ideas. In the language of the "universal characteristic," abstract logical theses would appear in the form of intuitively clear rules for operations with symbols. These rules would describe formal properties of transformations of symbols and be based on procedures that produce clarity in representation.

The universal characteristic, according to Leibniz, would have to be the source of a true logical algebra that would be applicable to various types of thought and would be the successor of Scholastic logic, an unfortunate victim of circumstances. In his view, the lack of success of Scholastic logic was due primarily to the lack of a rigorous and precise language subject to the rules of a carefully constructed formalization. The universal characteristic, however, would denote all simple elements in logical considerations by letters; complex logical considerations, by formulas; and sentences, by equations. This characteristic would make it possible to obtain all logical consequences that necessarily follow from observed data.

Sometimes, Leibniz used the French phrase *spécieuse générale* as a synonym for logical pasigraphy; it is natural to draw a parallel between this phrase and the phrase *analysis speciosa* of the French mathematician Viète (1540–1603), who used letters as symbols to denote quantities. The phrase *spécieuse générale* is reminiscent of Leibniz' aim of developing an adequate and universal system of notation that could be used to separate the contents of ideas into component elements by way of appropriate computation. This leads to the *combinatoria characteristica*.

Here is what Leibniz himself has to say about the purpose and problems of this "combinatorial art": "For me, the combinatorial art is the science or, as we might say, the characteristic or art of notation that deals with the forms or formulas of things in general, that is, their qualities in general or the relation between the similar and dissimilar in them; thus, for example, from given elements *a*, *b*, *c*, and so on, which may represent quantity or any other type of property, we may, by taking combinations, obtain very different formulas. This distinguishes *combinatoria characteristica* from algebra, which deals with quantitative formulas or the relations between equals and unequals. Algebra is therefore subject to the combinatorial art and constantly uses its rules. However, the rules of the combinatorial art are much more general and have applications not only in algebra but in decipherment, in various types of games, and even in (constructive) geometry as practiced in classical times; in brief, in all cases in which some similarity relation is under consideration" (cited in Reference 110, p. 190).

The second source of Leibniz' methodological ideas was the gnosiology of Giordano Bruno, who always placed great value on a symbolism that would be convenient in the mnemonic respect; he was the founder of monadology.

According to Bruno, the basic unit of being is the monad (see his work *De monade, numero et figuro*, 1591). The monad is the individual element of being in which matter and form organically mix. According to Bruno, the perfection of the whole and its parts results from the following considerations: the higher is implied in the lower; causes consist in actions; similarities consist in the *individuum*; and the very process of improvement is only a realization of potential possession. And it is in this thesis of infinite variety in the individual object that we can see the beginnings of Leibniz' monadology and, in particular, his theory of the analytic nature of true statements. Bruno attempted to find a system of signs that would be convenient in the mnemonic respect and attempted to find formal criteria for establishing the differences between useful and useless combinations of terms. He also put the *Ars Magna* on an epistemological foundation. Because there are three fundamental operations of reason (1) recognition of simple objects, (2) abstractions that generalize and particularize, and (3) proof, the methods of the great art must be comprised of the following three phases: (1) indication of initial terms (Lully's alphabet); (2) study of the methods of combining and separating terms; (3) proof on the basis of the corresponding rules for combining terms. Bruno devoted his attention primarily to the problem of "multiplication of terms."

Leibniz attempted to reconcile what he knew of Bruno's ideas about monadology with the traditional forms of nominalism. According to Leibniz, essences exist not in the form of universals but in the form of monads. To some extent, Leibniz' monads were a reaction against the thesis that the world consists of objects related by definite rules and having absolutely fixed properties, just as a building is made up of individual bricks whose properties can be described and stated once and for all.

The third source of Leibniz' methodology was René Descartes' conception of a universal mathematics (*mathesis universalis*). Indeed, it was the mathematical works of Descartes that led to the introduction of literal algebra and the systematic use of variables. Engels evaluated the results of Descartes as follows: "The Cartesian *variable quantity* was a turning point in mathematics. It was a result of this discovery that *motion* and *dialectics* entered into the study of mathematics, and this necessitated the differential and integral calculus, which immediately appeared and which was then completed, and not constructed, by Newton and Leibniz" (*Dialectics of Nature* [Russian translation], 1959, p. 206).

According to Descartes, mathematics is the universal science of order and measure. Leibniz adopted this thesis, stating it in the terms of his logical studies. Couturat described the essence of universal mathematics in the sense of Leibniz as follows: "Leibniz noted that there exists a universal mathematics, from which all of the mathematical sciences draw their principles and most general theorems, and that this mathematics merges with logic or, at least, is an important part of it. There is no mere formal analogy, no mere parallelism between mathematics and logic, but an identity or at least a partial identity. Indeed, universal mathematics is, as we have noted, the general science of relations. Each relation, being characterized by some number of formal properties, can be used as the basis for a special theory with its own axioms and theorems and, in the last analysis, as the basis for a calculus for which these axioms are the rules; in brief, it can be the basis of some algebra. Together with classical algebra, which is only the logic of numbers and quantities with the single relation of equality, Leibniz considered other algebras based on the relations of similarity, coincidence, etc., and the algebra of identities and inclusions, which includes classical logic. And because these algebras were based on the formal properties of formulas and sentences, the universal mathematics becomes a true formal logic, the science of laws and general forms of thought" (Reference 147, pp. 317–318).

There is reason to take the source of Leibniz' universal mathematics to be his study of argumentation with respect to form (*argumens en forme*), which is a justification for the necessity of making logic mathematical. In this connection he said, "By an 'argument of form' I mean not only the Scholastic method of argumentation but any argument that is based on form and requires nothing at all in addition. Thus (to squander still another syllogistic construction to avoid repetition), even a properly constructed system of calculation, the algebraic calculus, is for me approximated by arguments of form, since the form of the considerations used in such a calculus is predetermined so that we are assured of correct considerations" (*Nouveaux Essais*, vol. 4, chap. 17).

Leibniz not only made logic mathematical but mathematics logical. For example, in his *Specimen calculi universalis* he attempted to give the concept of number a purely logical definition. However, his attempt to include all of mathematics in formal logic failed, a fact which, in the light of the recent results of Gödel on the inherent incompleteness of formalized arithmetic, is completely comprehensible.

One of the central points in the methodology of Leibniz is his *study of the analytic nature of any statements.* In any statement, he believed, the predicate is part of the content of the subject, a part that can be extracted from the subject analytically (he said: "*Praedicatum inest subjecto*"). According to Leibniz, "a fixed predicate or subsequent is contained in the subject or antecedent, and here lies the nature of truth in general. This is true with respect to any truth — assertive, universal or partial, necessary or contingent" (cited in Reference 165, pp. 518–519).

Leibniz interpreted the inclusion of a predicate in its subject by means of the example of the divisability relation for numbers; a truth corresponds to a ratio (proportion) in which the antecedent (subject) is larger than the consequent (predicate), so that the one contains the other. If the ratio is commensurable, the total measure of both terms can be determined directly by means of the Euclidean algorithm. With incommensurable ratios, it is necessary to consider infinite continued fractions whose successive values yield increasingly accurate expressions for the incommensurable ratios they represent. The study of the analytic nature of sentences influenced Leibniz' theory of proof and truth. The proof of a proposition is complete, according to Leibniz, only when analysis of the subject of the proposition shows that the subject contains the predicate. Accordingly, complete definition of a subject requires that the decomposition and analysis of the subject be complete. Thus, from the viewpoint of Leibniz, analysis of concepts is more difficult than analysis of truth. But even if it requires an infinite process, the analysis of truth is nonetheless possible. Analysis can be applied to both concepts and propositions. Analysis of concepts consists in definining them, while analysis of truths reduces to proving them. As a result, Leibniz attempted to replace all of Descartes' rules by the following two:

1. In a science it is not permissible to use undefined terms (except for a small number of undefinable terms).
2. A science cannot use unproved statements (except for axioms). However, even axioms can be proved. The use of axioms in one or another system is dictated, in brief, by practical expediency.

A proof is constructed, said Leibniz, by decomposition of the terms of the statement under proof. Thus, analysis of truth or a statement reduces to analysis of concepts, i.e., essentially, to definition.

In a discussion of the relationship between proof and the operations of analysis and synthesis, Leibniz said that a proof is constructed by means of definition (analysis) and the operation of combining these definitions (synthesis) (in letters to Conring, March 1, 1678 and March 19, 1678). To elucidate this process of using definitions in proofs, we will initially consider Leibniz' attempt to prove the arithmetic proposition $2 + 2 = 4$ logically. Initially, Leibniz used the following definitions of "two," "three," and "four": $2 = 1 + 1$ (first definition), $3 = 2 + 1$ (second definition), and $4 = 3 + 1$ (third definition).

His proof contains the following steps:

$$2 + 2 = 2 + 2 \tag{1}$$

(substitution and the law of identity),

$$2 + 2 = 2 + (1 + 1) \tag{2}$$

(by the first definition),

$$2 + 2 = (2 + 1) + 1 \tag{3}$$

(actually by a tacit agreement about distribution of parentheses),

$$2 + 2 = 3 + 1 \tag{4}$$

(by the second definition),

$$2 + 2 = 4 \tag{5}$$

(by the third definition).

This proof uses definitions and the principle of substitution of equals for equals. However, the proof is not absolutely correct since it implicitly uses the principle of associativity of addition, $(a + b) + c = a + (b + c)$, that is, a proposition that is necessary for the proof.

We present excerpts from Leibniz' paper *Non Inelegans Specimen*

Demonstrandi in Abstractis, which gives an excellent idea of the role definitions play in the process of proof.

Definition 1. Two terms are the same if one can be substituted for the other without altering the truth of any statement: $A = B$ denotes that A and B are identical.

Definition 2. Terms which are not the same, that is, terms which cannot always be substituted for one another, are different: $A \neq B$ signifies that A and B are different.

Proposition 1. If $A = B$, then also $B = A$. For since $A = B$ (by hypothesis), it follows (by Definition 1) that, in the statement $A = B$ (true by hypothesis), B can be substituted for A and A for B; hence we have $B = A$.

Proposition 2. If $A \neq B$, then also $B \neq A$. Otherwise, we would have $B = A$ and, in consequence (by the preceding proposition), $A = B$, contrary to hypothesis.

Theorem 3. If $A = B$ and $B = C$, then $A = C$. For if, in the statement $A = B$ (true by hypothesis), C be substituted for B (by Definition 1, since $A = B$), the resulting proposition will be true.

Theorem 4. If $A = B$ and $B \neq C$, then $A \neq C$. For if, in the proposition $B \neq C$ (true by hypothesis), A be substituted for B, we have (by Definition 1, since $A = B$), the true proposition $A \neq C$.

Definition 3. To say A is in L, or L contains A, is the same as to say that L can be made to coincide with a plurality of terms, taken together, of which A is one; $B \oplus N = L$ signifies that B is in L; and that B and N together compose or constitute L. The same thing holds for a larger number of terms.

In the above text we have replaced Leibniz' sign ∞ for the equality relation by the more modern symbol $=$ (Reference 168, p. 234). Leibniz further developed the third definition (of the concept of inclusion) in the following scholium: "We say that the concept of the genus *is in* the concept of the species; the individuals of the species amongst (*in*) the individuals of the genus; a part in the whole; and indeed the ultimate and indivisible in the continuous, as a point is in a line, although a point is not part of a line. Likewise, the concept of the attribute or predicate is in the concept of the subject. And, in general, this conception is of the widest application . . .. We are not concerned here with the notion of 'contained' in general — i.e., with the manner in which those things which are 'in' are related to one another and to that which contains them" (Reference 168, pp. 234–235).

According to Leibniz, all truths are logically provable, except for various propositions concerning identities and empirical propositions based on experimental evidence. He later refined his theory of proof, distinguishing the following aspects of the process of proof: (1) definition, (2) axioms and postulates, (3) previously proved theorems, and (4) empirical truths. Rational truths cannot depend on experiment, and they reduce to definitions and the principle of identity.

Leibniz considered problems in the theory of defining concepts in a letter to the mathematician Chiringhausen (end of May, 1678) in which he indicated that the mark of a real definition is possibility. Namely, a real definition states that a substance is possible or exists. In order to make it possible to use a given definition, it is first necessary to prove that the object of the definition is possible, that is, the assumption of the existence of the object does not lead to a contradiction. According to Leibniz, a definition must include all conditions necessary and sufficient for the proof of all properties of the object being defined.

Abandoning the nominalist theory of Hobbes, according to which any definition is nominal, Leibniz treated nominal definitions as those that state a distinctive characteristic of the antecedent (subject) of a sentence. Thus, nominal definitions must be separative propositions. He also abandoned the Cartesian criterion of truth (a proposition is true if it is clear and obvious). Where, asked Leibniz, can we find a reliable criterion for truth?

In his opinion, any noncontradictory idea is possible. Hence, a necessary condition for truth is consistency, and it can be observed by decomposing the concept in question into simple elements. The method of making such observations is provided by the "universal characteristic."

The essential premise for the "universal characteristic" of Leibniz is the thesis that it is possible to find and systematically describe both simple and complex concepts, with the latter (which cannot be decomposed further) being the elements of thought. He first explicitly stated this idea in 1666 in his *Ars combinatoria.*

This philosopher attempted to create his own kind of true logical algebra. A logical calculus must provide not only a means for constructing logical proofs of statements already known but also a means of finding new correct results.

The theory of identities as developed by Leibniz occupies an important place in his arsenal of methodological conceptions and has a number of mathematical and logical applications.

The meaning of the concept of identity, he said, reduces to the possibility of equating objects which it is impossible to find reason to call different. The notion of identity therefore reduces to the concept of indistinguishability. By applying the so-called principle of identity to the differential calculus, Leibniz found that it was possible to consider the equations of this calculus as a special type of vanishing inequality. He tended to define the concept of identities by means of such notions as "has the property," "object," "each," and "distinction." Using the symbol $\sim$ to denote identity, we can represent the Leibnizian notion of identity by the expression

$$(x \sim y)_{Df} = [\forall F(F(x) = F(y))],$$

where $\forall$ is the universal quantifier for the arbitrary predicate F. In words, X is identical to Y if and only if anything that can be said about X can also be said about Y. In other words, X has every property belonging to Y, and Y has every property belonging to X. Sometimes an attempt is made to state Leibniz' law of identity in a form that does not use quantifiers.

Leibniz did not accept induction as a means of proving apodictic statements. Induction is the opposite of deduction and bears the same relation to deduction as integration to differentiation.

Concerning probabilistic logic ("on estimating uncertainty"), Leibniz used the following concepts: a continuous scale of probabilities, the principle of indifference ("equivalent hypotheses can be used equally well in computations" — this is one of the corollaries of the law of sufficient reason), the definition of probability as a measure of knowledge, and certain ideas about operations on probabilities.

The development of ideas in Leibniz' probabilistic logic was, to a large extent, due to the work of Jakob Bernoulli, who in 1713 explained to Leibniz that the probability of an empirical law increases with an increase in the number of experiments which verify it and that the combined influence of a very large group of random factors leads to results that are almost independent of individual cases. Leibniz distinguished a forward and reverse calculus of probabilities: the first calculus is concerned with estimating the probability of an occurrence if the probability of its causes or its conditions is known. The second considers methods of estimating the probability of causes when the probability of their influence is known. Two other important methodological principles of Leibniz were his *law of continuity* and his thesis concerning *the necessity of sufficient reason.*

The law of continuity establishes a certain continuity among the monads, and it is assumed that they comprise such an infinitely dense series that it is possible to pass from the properties of one monad to the properties of another without a jump. It is easy to see that the principle of continuity is, in essence, based on the Leibnizian analysis of the infinitely small, in particular his concept of a real, infinitesimal quantity. From the mathematical viewpoint, Leibniz' principle of continuity describes the concept of the continuous variation of a function, which is a necessary, although not sufficient, condition for differentiability. From the methodological viewpoint, the law of continuity affects problems of passing from one qualitative state of an object to another by introducing a scale of degrees of distinction between them and reducing this distinction to a minimal level. The notion of a real, infinitesimal quantity must, according to Leibniz, be used to express continuity in the process of passing from one qualitative state to another.

The law of sufficient reason appears as a methodological postulate that Leibniz intended to use in the natural sciences, where the role of the principle of contradiction is minimal. According to Leibniz, "the principle of the necessity of sufficient reason consists in the fact that nothing occurs without sufficient reason for it to occur rather than something else" (Reference 172, vol. 8, p. 355). As a good example of the meaning of the postulate of sufficient reason, Leibniz presents the Archimedean lever (equal weights located at equal distances from the fulcrum of the lever are in a state of equilibrium). A uniform lever does not move, because there is no sufficient reason for it to leave the state of equilibrium.

Leibniz was inclined to use the principle of sufficient reason outside the realm of logic. Logical laws, he thought, were laws for all possible worlds, while the law of sufficient reason concerns the real world and provides an interpretation of it to be used for discovery of certain purely physical principles. According to Leibniz, the law prohibiting formal contradictions determines logical and mathematical truths, while the truth of physical and moral statements may compete with the law of sufficient reason. In this connection he wrote, "the principle of finding sufficient reason is unnecessary in arithmetic and geometry, but it is absolutely necessary in physics and mechanics" (Reference 172, vol. 8, p. 27).

From the ontological viewpoint, said Leibniz, application of the law of sufficient reason is based on the following axioms: equilibrium, the principle of indistinguishables, and the principle of the simplicity of natural laws.

There is still one more aspect of the law of sufficient reason that we should note. This law must, according to Leibniz, provide the basis for proofs of contingent (empirical) truths since it is impossible to prove them directly. Leibniz gave special consideration to the relationship between the law of sufficient reason and the law of identity, which asserts that the truth of any statement of a tautological character reduces to a proof that any true statement is analytic, that is, is identically possible. Leibniz was also concerned with the analysis of the law of the excluded middle. For example, he treats the law $x \supset \bar{\bar{x}}$ as a conversion of the law of the excluded middle.

Leibniz brilliantly applied his methodological conceptions not only to theoretical studies but to practical matters as well. In its utilitarian aspect, he insisted, logic must actively advance mathematics, physics, geology, philology, jurisprudence, and technology. The value of logical proof (by analogy with geometric proof) is great even in the natural sciences (see Leibniz' letter of the end of October 1671 to Johann Friedrich).

Concerning practical applications of his combinatorial art, which he thought must play a role in the generalized theory of relations and orders, Leibniz had the following to say: "In mathematics and mechanics I found, by means of the combinational art, various things having no small value in practical life: first, I found an arithmetic machine, to which I gave the name 'living account book' . . . and which has applications in human affairs, the military arts, offices, sine tables, and in astronomy" (cited in Reference 100, p. 569).

An attempt to bring the methods of mathematics and physics closer to each other was characteristic of Leibniz' methodology. The possibility of doing so, he thought, was dictated by the fact that both use essentially deductive methods of research. The differences between them are no greater than those between synthetic (progressive) and analytic (regressive) methods used in natural science. The regressive method is essentially related to the use of a hypothesis, and Leibniz attempted to study the degree of probability of hypothetical assumptions. He reasoned that the probability a hypothesis is true increases as its simplicity increases, and the larger the group of phenomena covered by a given hypothesis with the smallest possible number of postulates, the more probable the validity of the hypothesis. According to Leibniz, logical analysis of hypotheses is a necessary condition for construction of "a logic of discovery of new truths" or an "art" of finding algorithms.

It is impossible to conclude a survey of Leibniz' methodology without at least a brief discussion of how it was affected by his theory of monads.

It is hardly possible that the content of Leibnizian monadology can be exhausted by its theological aspects and these aspects alone. To understand this theory, we must recall Leibniz' treatment of the concept of kinetic energy which reduces to the equation

$$\sum_{i=1}^{i=n} m_i v_i^2 = \text{const}, \tag{1}$$

where $n \to \infty$, $\sum$ is the summation sign, m is the mass of an object, and v is its velocity. Leibniz attempted to extend Equation 1 not only to material things but even to the spiritual activity of mankind. Recognizing that application of Equation 1 to rigid bodies meets with serious difficulties, the philosopher attempted to lay a teleological basis under his hypothesis about the pseudoloss of kinetic energy in such cases (it is conserved, said Leibniz, in molecular vibrations). Here reference is made to the postulate of the so-called pre-established harmony, which leads to extensive considerations concerning the desirability of the structure of the universe, and so on. Thus, under the outer layer of teleological terminology (which is, in essence, merely an evasion), Leibniz' view leads to an approach that is methodologically consistent with the modern law of conservation of energy, which holds that energy is conserved in such reactions by conversion from one form into another. We can corroborate this interpretation by referring to what V. I. Lenin had to say: In his monadology, Leibniz "theologically approached the principle of the continuous (and universal, absolute) relationship between matter and motion" (*Filosofskiye Tetradi*, Moscow, 1947, p. 313).

4. Leibniz' Logical Results; Three Stages in the Development of His Logical Calculus

The importance of logic in Leibniz' system is due, in the opinion of Russell, to the fact that "Leibniz, in his private thinking, is the best example of a philosopher who uses logic as a key to metaphysics" (Reference 164, p. 613). However, this view is one-sided in the sense that for Leibniz philosophical conceptions also had an effect on his logical views. For example, his analytic theory of statements can be treated as one of the corollaries of his monadology.

According to Leibniz, logic is the science "that teaches other sciences a method for discovering and proving all corollaries following from given premises" (Reference 235, vol. 1, p. 168). In his works he sometimes uses the terms *Vernunftkunst* and *Denkkunst* for logic.

The fundamental principles of Leibnizian logic can be stated briefly as follows:

1. Any concept can be reduced to a fixed set of simple concepts (that is, concepts that cannot be further decomposed). This set of concepts is chosen from some collection of elements forming an "alphabet of thought."
2. Complex concepts can be derived from simple concepts only by means of the operation of logical multiplication, which corresponds to conjunction in propositional calculus and the operation of intersecting volumes of concepts in the logic of classes.
3. The set of elementary concepts must satisfy the criterion of consistency.
4. Any statement is a predicate in the sense that it can be translated into an equivalent form in which the predicate is already part of the subject.
5. Any true affirmative proposition is analytic in the sense that its predicate is contained in the subject (which is natural for the case of an intensional interpretation).

It is not difficult to see that these ideas of Leibniz are closely related to his methodology. For example, the fifth thesis is only a logical statement of the Leibnizian notion of the law of sufficient reason. Just as a subject provides a sufficient condition for the predicate of a sentence, a cause is a sufficient reason for an action.

Leibniz made three clearly differentiated attempts at constructing a logical calculus, the first prior to 1679, one in 1685–1686, and his last in 1690.

We will first consider the logical results he obtained in 1679 in his papers *Specimen calculi universalis*, *Specimen calculi universalis addenda*, *Elementa characteristicae universalis*, and *Elementa calculi*. Here he uses prime numbers to symbolize elementary concepts. Complex concepts are represented in the form of products of elementary concepts, and a numerical characteristic is associated with each concept. Assume that the number 6 corresponds to the concept of "human," that the number 3 corresponds to the concept of "animal," and that the number 2 corresponds to the concept of "reasoning"; in this case the equation $6 = 2 \cdot 3$ corresponds to the definition of the concept of "human."

Leibniz wrote the universally affirmative proposition (all S are P) of syllogistics in the form $S = P \cdot y$, which, if we use the existential quantifier, can be written in contemporary notation as follows:

$$\exists y(S = P \cdot y), \tag{1}$$

where $\exists$ denotes the existential quantifier. Leibniz represented partially affirmative propositions (there exist S that are P) in the form $S \cdot x = P \cdot y$, which can be rewritten in the form

$$\exists x \exists y(Sx = Py). \tag{2}$$

Universally negative propositions (no S are P) can be written in the form

$$\forall x \forall y(Sx \neq Py), \tag{3}$$

where $\forall$ denotes the universal quantifier. Finally, partially negative propositions (there exist S that are not P) admit the representation

$$\forall y(S \neq P \cdot y). \tag{4}$$

Application of Representations 1 through 4 to a symbolization of concrete sentences proved to be fraught with great difficulties. Indeed, according to 3, universally affirmative propositions must be true if $\overline{\exists x \exists y}(\overline{Sx} = Py)$. However, it is always possible to choose x and y so that $Sx = Py$; that is, we find that there is no true universally negative proposition.

Leibniz evaded difficulties by introducing two numbers for each characteristic — one with a plus sign, and the other with a minus. In contrast to ordinary arithmetic, Leibniz set all divisors of a positive numerical characteristic equal to positive numbers and all divisors of negative characteristics equal to negative numbers. Under these circumstances, we can express the assertion that one notion is contained in another by requiring that both numerical characteristics of the included concept be divisors of the corresponding characteristics of the containing concept.

Leibniz now based his arithmetic interpretation on the correspondence between syllogistic variables (terms) and ordered pairs of relatively prime natural numbers (that is, numbers having no positive common divisors except for unity). Let two relatively prime numbers R_1 and R_2 correspond to a variable, S, and two other relatively prime numbers T_1 and T_2 correspond to another variable, P. Then the proposition "all S are P" is true if and only if R_1 is divisible by T_1 and R_2 is divisible by T_2. If one of these two conditions is not satisfied, the proposition "all S are P" is false. The sentence "some S are P" is true if and only if R_1 and T_2 are relatively prime and, likewise, R_2 and T_1 are relatively prime. If at least one of these conditions is not satisfied, the proposition "some S are P" is false. Leibniz selected several modes and rules of syllogistics at random and verified them by means of his interpretation.

Leibniz also stated a series of theorems concerning propositional calculus and the logic of classes. He used the word *est* as a symbol for the copula "is" and expressed identity by means of the phrases *eadem sunt* and *sunt idem*. Leibniz clearly distinguished axioms (*propositiones per se verae*, that is, propositions true in themselves) and propositions that could be derived from them within his system (*verae propositiones*).

Among the logical theorems formulated by Leibniz in 1679, we note the following:

1. *A est BC* if and only if *A est B* and *A est C*.
2. If *A est B* and *C est D*, then *AC est BD* (here and in the following we indicate conjunction by omitting symbols between letters).
3. *AB est A*.
4. *AB est B*.

Leibniz applied Theorem 2 to the case where A and C, as well as B and D, are, respectively, incompatible concepts. This is indicated by the following text, in which he gives a concrete example of Theorem 2: "A circle is a figure with no angles. A square is a quadrangle. Consequently, a circle that is a square is a plane figure with no angles that is also a quadrangle. The conclusion is a true proposition consisting of two other propositions that are each impossible" ("Circulus est nullangulus. Quadratum est quadrangulum. Ergo circulus — quadratum est nullangulus — quadrangulum. Nam heac propositio vera est ex hypothesi impossibili," *Specimen demonstrandi*; cited in Reference 166, p. 99).

The works of the period under discussion contain the following forms of four syllogistic statements: *S est P* (universally affirmative), *S est non P* (universally negative), *S non est non P* (partially affirmative), *S non est P* (partially negative). In Expressions 1 through 4 Leibniz sometimes used the symbols *A*, *B*, *C*, and *D* to denote properties (with the grammatical analogs of participles and adjectives), and sometimes he used them to denote classes of objects (here the grammatical analog is the noun). In the first case the particle *est* is used to indicate that the subject has a given property, while in the second case it is used to denote the relationship of inclusion of a smaller volume in a larger one. The intensional interpretation (first case) is isomorphic to the extensional interpretation (second case).

The second stage in the development of Leibniz' logical system occurred the same time his philosophy matured, that is, during the period 1685–1686. The following papers, in which Leibniz presented an algebraic exposition and development of logic, contain material characteristic of this period: *Generales inquisitiones de analysi notionum et veritatum*, *Principia calculi rationalis*, and *Difficultates quaedam logicae*.

Here Leibniz used the symbol ∞ for the copula and, in words, the expressions *coincidunt* and *aequivalent*. He also introduced the term *Ens* (or *Res*), which corresponds, in the intensional interpretation, to the null class and, in the extensional treatment, to the universal class. The Leibnizian relations

1. *S est P aequivalent S non — P est non — Ens*,
2. *S est non — P aequivalent SP est non — Ens*,
3. *S non est non — P aequivalent SP est Ens*,
4. *S non est P aequivalent S non — P est Ens*

have the following forms in contemporary notation:

1′. $SaP \rightleftarrows (S\bar{P} = 0)$,

2′. $SeP \rightleftarrows (SP = 0)$,

3′. $SiP \rightleftarrows (SP \neq 0)$,

4′. $SoP \rightleftarrows (S\bar{P} \neq 0)$,

where 0 denotes the empty class, $\rightleftarrows$ denotes equivalence, $\neq$ denotes inequality, and *S* and *P* are taken to be classes of objects.

Here is the corresponding text: "Any *A* is *B* . . . or (*seu*) *A* that are not *B* do not exist (*est non — ens*). Some *A* are not *B* . . . or *A* that are not *B* exist. No *A* is *B* is equivalent to *AB* does not exist. Some *A* are *B* is equivalent to *AB* exists" (see Reference 166, p. 102).

This translation of classical statements into the form of existential statements was, however, later eliminated by Leibniz. He could not use this form to cope with the difficulty encountered in studying the problem of converting sentences. Unfortunately, Leibniz began with the improper assumption that all of Aristotelian logic must also be derivable in his calculus. As a result, he had to treat general propositions as those that imply the existence of their subjects, which partially explains some of the inconsistency in his development of an axiomatic structure for logic.

Leibniz formally defined the operation of logical multiplication as follows: If *A est B* and *A est C*, then *A est BC*. He actually defined inclusion in terms of equality, assuming that the expression *S est P* is equivalent to $S \infty SP$. Thus (S *est* P) *est* ($S \infty SP$). Setting $S \infty P$, we have (S *est* S) *est* ($S \infty SS$), but the first premise can be eliminated, because it is always true (the law of identity). From this it follows that

$$S \infty SS.$$

Thus, Leibniz discovered the law of idempotence (here, the symbol ∞ denotes equality).

Using the copula *est*, Leibniz clearly distinguished between its extensional interpretation (here *A est B* means that *A* is contained in *B* in volume) from its use in the intensional sense (here *A est B* means that *A* contains *B* in meaning). In the latter case (the intensional treatment) he wrote "*A continet B*" (*A* contains *B*) instead of "*A est B*."

As Couturat (Reference 147) emphasized, Leibniz discovered a completely different interpretation of his second logical system. In Rescher's opinion (Reference 123), this interpretation can be treated as a predecessor of Lewisonian strict implication. The variables A, B, C, $D, \ldots$, according to Leibniz, must here be treated as denoting propositions to be taken as indivisible entities, *non* must denote negation of propositions, a series of letters (without symbols between them) must be interpreted in the sense of conjunction; *est* denotes the relation of implication with respect to content; *ens* must be taken as a logical necessity (a logical truth); and correctness must be identified with logical consistency ("*Cum dico A est B, et A et B sunt propositiones, intellego ex A sequi B*," as Leibniz said).

In *Generales inquisitiones de analysi notionum et veritatum* this interpretation was properly qualified as a modal logic. Leibniz also defined implication in terms of negation, conjunction, and the notion of possibility (Reference 147, p. 355).

The next and last attempt at expanding his logical system was made by Leibniz in 1690. His works *Fundamenta calculi logici* and *De formae logicae comprobatione per linearum dictus* belong to this period.

Rescher (Reference 123) considered Leibniz' third logical system an attempt at axiomatizing the theory of the notion of "inclusion." Leibniz translated the expression "contained in" by the word *inest*. In addition to the binary relation *inest*, he also introduced two other binary relations: (1) noncommunication (*incommunicantia*) and (2) its negation, communication (*communicantia*). We will denote noncommunication with the symbol $|$ and communication with the symbol $\uparrow$. Leibniz used the symbol $+$ to denote indivisible disjunction.

We should note the following theorems and definitions of the third system:

1. if A *inest* B and C *inest* B, then $A + C$ *inest* B;
2. if A *inest* B, then $C + A$ *inest* $C + B$;
3. if A *inest* B and C *inest* D, then $A + C$ *inest* $B + D$;
4. $A - B = C$ if and only if $A = B + C$ and $C|B$.

The operation of logical subtraction is introduced in 4. Note that the subtrahend may not exceed the minuend, a condition which, in contrast to the case of algebra, eliminates negative quantities from logic. The indeterminacy in the subtraction operation, as well as the fact that it is not defined for all values of the variables (for example, it becomes meaningless if, in $A - B$, we set A equal to a false statement and B equal to a true statement) was a point of concern for Leibniz in his attempt to find the logical analog for every algebraic operation. He did not even attempt to eliminate subtraction by expressing it in terms of other logical functions under additional constraints.

The difficulties associated with the treatment of logical subtraction constituted an additional influence that prevented Leibniz from finding satisfactory forms of logical calculus.

Some of the relations in the last variant of Leibniz' calculus read as follows.

5. $A - A = N$, where N denotes a contradictory property;
6. $(A - B) \mid B$;
7. $A \uparrow B$ if and only if there exists a C different from A such that C *inest* A and C *inest* B;
8. $A + N = A$;
9. N *inest* A.

Theorem 7 defines the notion of communication in terms of "inclusion" and "contradictory property," whose formal properties are described in 5, 8, and 9.

There is reason to agree partially with Rescher in his statement that "if Leibniz' logical calculus did not reach the perfection and elegance of later logical algebras, it was not because of his intensional notion of logic but because of his adherence to traditional logic, which tended to rest on narrower logical considerations rather than broader algebraic considerations" (Reference 123, p. 13). To this we might add the absence of broad stimuli from the natural sciences for development of such a calculus during Leibniz' era, as well as the fact that using the inverse operations (subtraction and division) introduced by Leibniz engenders considerable difficulty of a formal nature. In conclusion, we should note Leibniz' contribution to the development of several problems in traditional logic that were considered in various places in his *New Essays*. He noted the profound analogy between categorical and hypothetical statements. Even before Euler, he used circles to illustrate the volumes of notions, and lines for geometric illustration of the relations between the various terms of sentences.

Leibniz augmented the proposition *SaP* by the condition "*S* is possible," and he added the following five modes to the traditional list of modes of figures of categorical syllogisms: *Barbari*, *Celaront* (first figure), *Cesaro*, *Camestros* (second figure), and *Calemos* (fourth figure); these are additions because, from the viewpoint of syllogistics, it is impossible to justify the validity of the conclusions $a \rightarrow i$ and $e \rightarrow o$.

Leibniz devoted a special investigation to the role of identical expressions in syllogistic proofs. "I assert," he said, "that the one principle of contradiction is sufficient to prove the second and third figures of a syllogism, given the first" (Reference 174, p. 320). Leibniz used an argument of the following type to prove how *Bocardo* can be obtained from *Barbara*. The conclusion *Barbara* has the form $((BaC) \& (AaB)) \rightarrow AaC$, where & denotes conjunction and the arrow denotes implication. Assume that *AaC* is false, that is, that *AoC* is true. Then either *BaC* or *AaB* is false. Assume that *AaB* is true, so that *BaC* is false; that is, we have *BoC*. We now make *BoC* the conclusion of a new syllogism whose first premise is *AoC* (the contradictory of *AaC*, which is false by hypothesis) and whose second premise is the old *AaB*. The new syllogism reads as follows:

$$((AoC) \& (AaB)) \rightarrow BoC.$$

This is *Bocardo* of the third figure.

There is some reason for asserting that Leibniz sensed the narrowness of the Aristotelian theory of propositions, which attempted to reduce all forms of statements to the form *X* is (is not) *Y*, where *X* is the name of the subject and *Y* is the predicate. From this point of view, in fact, how can one proceed to a logical description of relations? Leibniz also explicitly deviated from the Aristotelian approach when he characterized the notion of relations between two arbitrary objects *L* and *M* as follows:

"We cannot assert," wrote Leibniz, "that *L* and *M* together both comprise the subject of an attribute, since we would then be forced to deal with attributes that, as it were, had one foot in one subject, and the other in the other subject, which is incompatible with the notion of attributes. Accordingly, we must say that a relation . . . is actually outside of subjects" (fifth letter to Clark). Here he explicitly attempted to replace the classical formula of an attributive statement with the relational statement xRy of the logic of relations, where, in the words of Leibniz, the relation R is "outside the subject" x and y.

The intuitive dissatisfaction Leibniz felt about Aristotelian logic appeared in his attempts to find nonsyllogistic forms of propositions. Among these forms, he considered inferences of the type "x is the uncle of y, so y is the nephew of x." In modern terms, Leibniz concludes from the truth of the relation $R(x, y)$ the truth of the converse $\breve{R}(y, x)$. The implication $R(x, y) \rightarrow \breve{R}(y, x)$ is not explicitly considered in the Aristotelian catalog of forms of propositions. In fact, as professor P. S. Popov properly notes, Leibniz, in his study of Descartes' system, stated the following problem: Is the first principle of rationalism, "I think, therefore I am," a proposition? If so, it cannot be taken as an initial fact. If it is a first principle, what does the word "therefore" mean? Descartes was not presenting any propositions in his first principle. Concerning this, Leibniz noted that the proposition "I exist" is obvious since it cannot be proved on the basis of any other propositions — it is an immediate truth. To say "I think, therefore I exist" does not prove existence from thought since to think and to be thinking are the same thing, and to say "I am thinking" is the same as saying "I am, I exist." Here Leibniz again used the principle of identity and, following Descartes, said that there is no basis for attempting to find a syllogism in this first principle (Reference 109, p. 97).

Leibniz demonstrated that Peter Ramus' conjecture about the probability of the classical operation of conversion of propositions by means of the second and third figures of a categorical syllogism is true. He distinguished the following three types of conversions:

1. $AeB \rightarrow BeA$,
2. $AiB \rightarrow BiA$,
3. $AaB \rightarrow BiA$,

where conversion 3 is subject to the assumption that A is not N.

Leibniz proves conversion 1 by means of *Cesare* $((AeM)\ \&\ (BaM)) \rightarrow BeA$, setting $M = B$, which yields $((AeB)\ \&\ (BaB)) \rightarrow BeA$.

The statement BaB is eliminated as a law of logic (the principle of identity), leading to the assertion that conversion 1 is provable. Similarly, he proved 2 by means of *Datisi*, and 3 by means of *Darapti*. In these proofs the trivial law of identity finds a non-trivial application.

Following Jung, Leibniz also noted the existence of a class of inferences that is not taken into account in ordinary school logic, the class of so-called inferences drawn by restricting the volume of an idea (in the terminology of Leibniz himself, "*a recto de obliquum*"). For example, a proposition of this type is "a cow is an herbivorous animal; consequently, the udder of a cow is the udder of an herbivorous animal."

Leibniz directed his logical studies primarily toward creative development, but he was nonetheless quite familiar with the history of logic (as, incidently, he was with the history of philosophy). Leibniz was not yet fifteen when he "wandered for whole days in a grove to choose between Aristotle and Democritus" (Reference 172, vol. 3, p. 205). In particular, he was completely at home in Scholastic logic, recognizing both its definite advantages over Aristotelian logic and its weaknesses.

As Leibniz saw it, the advantages of Scholastic logic were that it broadened the horizons and realm of application of Aristotelian analytics, developed problems in mnemonics, found new forms of propositions, and so on. On the other hand, the basic shortcoming of Scholastic logic was that it could not generalize its results directly by application of appropriate general methods of investigation (methods of an algorithmic nature). It was not accidental, therefore, that Leibniz preferred to direct his attention to the deductive-axiomatic aspect of logical research, which led him to become the originator of the first logical algorithms.

Not all of Leibniz' logical program has withstood the test of time. In particular, the development of modern science has proved that the "universal characteristic" is unrealizable. This typically metaphysical attempt at reducing all of meaningful human thought to a finite number of formal mathematical calculuses proved to be unsound. Naturally, the consequences of this notion, such as Leibniz' attempt to restrict all of meaningful mathematics to the narrow frame of formal logic, were also doomed to failure. However, Leibniz' ideas of making logic algebraic and his dream of a computational treatment of problems in natural science have been brilliantly realized in the progressive development of modern science and practice.

CHAPTER THREE

The Development of Symbolic Logic After Leibniz; the Seventeenth and Eighteenth Centuries

1. The Logical Achievements of Leibniz' Disciples

The Swiss mathematicians Johann Bernoulli (1667–1748) and Jakob Bernoulli (1654–1705), as disciples of Leibniz, were completely familiar with their master's attempts at finding satisfactory logical algorithms.

The brothers Bernoulli belonged to a Protestant family that had escaped the religious persecution of Duke Alva of the Netherlands by fleeing to Basel.

At the insistence of his father, Jakob Bernoulli followed a theological course of study, concurrently studying mathematics secretly. In 1676 he found himself in Bordeaux, where he earned a living by giving lessons in mathematics. In 1681 he returned to Basel, where he published an article on the theory of comets. In an article in *Acta Eruditorum* (1686), he explicitly used a very important method of mathematical argument — the method of complete mathematical induction.

In 1687 Bernoulli was appointed professor of mathematics at Basel, where he lectured on experimental physics, simultaneously studying combinatorial problems thoroughly and fruitfully. He first studied combinations, and in his writings he used the term "permutation" in the same sense as it is now used in algebra. He also analyzed permutations with repetitions and introduced the name "integral" for Leibniz' symbol $\int$.

Jakob Bernoulli is best known for his work in the theory of

probability, to which he devoted his *Ars Conjectandi* (1713); he distinguished *a priori* and *a posteriori* probabilities (that is, in modern terms, logical and empirical probabilities) and discovered the law of large numbers. Ironically, Bernoulli died just when he began to write the fourth part of *Ars Conjectandi*, the section concerning application of probability theory to the problem of human longevity.

Johann Bernoulli, Leonard Euler's teacher, initially chose a mercantile career but, under the influence of his brother, rapidly switched to mathematics and medicine. In 1695, on the recommendation of Huygens, Johann was appointed professor of physics at Groningen. In 1705 Johann was appointed Professor of Greek and Mathematics at Basel.

Johann Bernoulli stated the mathematically important brachistochrone problem, which stimulated further development of mathematical analysis. In *Acta Eruditorum* he presented a method for constructing certain first-order differential equations, and in *Journal de Scavants* (1697) he stated the problem of finding shortest paths on convex surfaces.

The Bernoulli brothers apparently occupied themselves with problems in logic until their relationship became acutely strained. In 1685 they noted a parallelism between certain logical and algebraic operations and presented their observations in their work *Parallelism between Logical and Algebraic Arguments* (Basel, 1685) (see References 177 and 178).

"For the notation of the operation of adding ideas of several substances, without affirmation or negation," they wrote, "we use the sign $\mathscr{E}$, as, for example, the set $\mathscr{E}$ of eruditions; the operation of adding ideas of several quantities that are not comparable is expressed by the sign +, as, for example, $a + b$. If, from an idea of larger order, we separate an idea of lesser order, the residue of the first (difference) is formed; thus, in the notion of "man," in addition to "animal," there is the notion of "reasoning," so that, by eliminating the notion of "animal" (from the notion of "man"), we are left with "reasoning" as the difference. In like manner, if we subtract a smaller quantity from a larger, we are left with the difference between the one and the other; this is expressed by the symbol −, as, for example, $a - b$ denotes the difference between

a and b. When two ideas between which the intellect considers there to be agreement, identity, inconsistency, or contradiction are simultaneously affirmed or denied, then, as a notation of these relations in words, we use "*is*" or "*is not*." The process of establishing the relations under discussion is called expression (*enunciatio*), as, for example, "man is an animal; man is not a nonreasoning being." Two quantities between which the intellect considers there to be equality are related by the sign of equality $=$ in an equation as, for example, $a = b$. On the other hand, inequality is denoted by the symbols $<$ and $>$, as, for example, $a < c$ or $a > c$ (Reference 177, cited in Reference 178, p. 125).

Here logical operations are treated intensionally. The operation of logical subtraction is not defined for all values of variables (it is impossible, in particular, to separate a higher-order idea from a lower-order idea; that is, in the expression $x - y$ the term y cannot exceed the term x). An expression that has no logical interpretation is the difference $0 - 1$, where 0 denotes falsity, and 1 denotes truth (the Bernoulli brothers did not take $x - y$ to mean $x \,\&\, \bar{y}$). As a result, they did not take the operation of logical subtraction to be a function of truth in the same way as, for example, the material implication of the Stoics. This complicated the algorithmic processing of those expressions which included the sign of logical subtraction. It is therefore not difficult to see that the Bernoulli brothers were unable to construct a satisfactory logical calculus. Following Leibniz, they only noted the analogy that exists between the laws of formal logic and the methods of elementary algebra. Nonetheless, it is the opinion of V. I. Glivenko (Reference 179) that the Bernoullis actually began the study of distributive structures long before Boole. Literature on the logical results of the Bernoullis may be found in References 179 through 182.

Another disciple of Leibniz, the well-known German philosopher Christian Wolff, deserves mention in the history of the early development of symbolic logic.

A follower and systematizer of the ideas of Leibniz, Wolff was born on January 24, 1679 into the family of a baker in Breslau. Wolff studied at the Breslau high school and was introduced by his teachers to the works of Descartes and the *Medicina mentis* (1689) of

Leibniz' correspondent Walter von Chiringhausen (1651–1708). In 1699 Wolff traveled to Vienna with the intention of studying theology, although he soon turned toward mathematics and philosophy. In 1702 he traveled to Leipzig, where he won his Master's Degree with his Latin thesis *On a Universal Philosophy Constructed with a Mathematical Method*, in which he follows Descartes in the most important points. In parallel with his study of Descartes, Wolff proceeded to serious work on Leibniz' writings.

In 1703 Wolff was appointed lecturer in Leipzig and, through the influence of Leibniz, professor in the department of mathematics at Halle (1706). Here he gave lectures on mathematics, physics, and all fields of philosophy. In the fields of metaphysics, morals, and law, Wolff stated fundamental ideas, of a moderate form, about the theory of natural law, as well as the abstract ideals of humanism.

Having exhausted "scientific" arguments, plotting theologians (among them the pietists Franke and Lange) who could not let Wolff's philosophical successes rest in peace decided to try more "efficacious" means — a scribbled denunciation of Wolff to King Friedrich Wilhelm I. Knowing the weak side of their ruler, the scheming theologians explained to the king that Wolff's ideas about pre-established harmony (and the associated thesis that free will does not exist) imply, in particular, that army deserters cannot be found guilty since their desertion was predetermined. The king was no master of Leibnizian metaphysics, but he did know a danger to the Prussian state. Thus, Wolff was forced to flee the royal wrath, escaping to Kassel in 1723, where he became the first professor in the philosophy department at Marburg. Among the students of Wolff at Marburg was M. V. Lomonosov.

Wolff gradually obtained worldwide recognition; he was made a member of the Academies in Paris (1733), London, and Stockholm. Peter I conferred upon Wolff the honor of Vice-Presidency of the Russian Academy of Sciences, whose opening was planned at the time for Petersburg. The worldwide fame of Wolff finally had a sobering effect on the Prussian court, and Friedrich II offered Wolff the position of Vice-President of the Academy of Sciences in Berlin. However, Wolff's cosmopolitan career predisposed him to return to Halle (1740), where he was met as a conquering hero. He died on April 9, 1754.

Wolff's works are cast in the deductive-geometric form of the *Ethics* of Spinoza. He attempted to obtain maximum clarity, although in metaphysics his style sometimes combines a deliberate deadness with pedantry. Kant considered Wolff the father of the spirit of dryness in German philosophy.

Philosophy, according to Wolff, is the general theory of possible substances, that is, objects that can be thought of in a non-contradictory manner (those objects about which the assumption of existence does not imply a contradiction). Ontology analyzes the foundations of universal consistent relations; mathematics is based on categories of measurement; and psychology splits into empirical psychology (in which the "soul" is investigated in close connection with the body) and rational psychology (in which such research is not permitted).

On the whole, Wolff's philosophical system was dualistic and eclectic. It included individual ideas of Spinoza, Descartes, and even Locke, whose influence appeared, in particular, in Wolff's energetic emphasis of the decisive value of experimental information in various branches of science (for example, physics). At the same time, Wolff emphasized the rationalistic method of demonstration in the deductive sciences. On the other hand, although he outlined the pattern of the universe as an analog of hours (of a clock), he explicitly followed Descartes and the French materialists.

Literature on the logic and philosophy of Wolff may be found in References 183 through 192 and 196 through 199.

We should note the following works of Wolff: *Rational Thoughts about the Powers of the Human Intellect* (St. Petersburg, 1765); *Philosophia rationalis sive Logica methodo scientifica pertractata et de usum scientiarum atque vitae aptata* (Halle and Leipzig, 1728); *Psychologia rationalis methodo scientifica pertractata . . ., Psychologica empirica . . .* (Leipzig, 1732); *Philosophia Wolfianae contractae tomus I: logica . . .* (Halle, 1744–1745); *Briefwechsel zwischen Leibniz und Chr. Wolff* (Halle, 1860).

Just as Wolff attempted to simplify the ideas of Leibniz in philosophy, he actually attempted to adapt those of Liebniz' views to which he was attracted. Like Leibniz, he emphasized the value of analytic propositions in mathematical and logical proofs. To Wolff, analytic sentences were "empty propositions" and were defined by him as "truths that serve as justifications for propositions in which the antecedent coincides with the consequent" (Reference 193, p. 91).

In a discussion on operations with aggregates of ideas, Wolff mentioned so-called conjugation of ideas (*Verknüpfung der Begriffe*), an operation akin to union of classes (without elimination of elements common to both classes).

Wolff devotes much attention to a discussion of the thesis of so-called "sufficient reason," which he was inclined to treat as a logical principle and which is also the starting point for his metaphysics in the following form: "'Nothing' is meaningless, so that all that exists must have sufficient reason" (Reference 194, p. 2).

Following Peter Ramus and the Cartesians of Porte Royale, Wolff freed logic from its Scholastic variant both in form and in content. In a somewhat simplified form, he attempted to account for Chirighausen's concept of logic.

The theoretical part of logic, according to Wolff, must describe ideas, statements, and conclusions. Its practical part must include problems on criteria of truth, degrees of probable considerations (here we can see the influence of Leibniz' probabilistic logic), the distinction between observation (*a posteriori*) and prediction (*a priori*), and finally the value of logical ideas in everyday life.

Wolff's statements of a number of fundamental logical ideas and principles were insufficiently clear. For example, his statement of the law of the excluded middle is, in fact, a principle combining the law of the excluded middle and the law of contradiction: "Of two contradictory statements, one must be true and the other false" (Reference 198, Section 532). Similarly, as we can easily see, Wolff's definition of conclusion ignores the existence of non-syllogistic conclusions: "Conclusion is the logical operation by means of which we can derive from two statements with a common term a third statement consisting of terms in which the initial statements differ from each other" (Reference 198, Section 50). If we agree with Wolff, every conclusion is syllogistic. Deviating somewhat from the mathematical practice of his time, Wolff attempted to proceed without generic definitions. Instead, he tried to establish "universal axioms" of mathematics; however, his axiomatization was not really an axiomatization and Hegel was completely justified in criticizing it. The reader can consult Reference 199 for further information on Wolff's "axiomatization."

Wolff's classification of reasons to be used in proofs is of interest; he considered the following as such reasons or premises (*Voraussetzungen*):

1. definitions or tautological premises, in which the subject and its predicate denote the same idea;
2. obvious observational data (*unbezweifelte Erfahrungen*) established by careful and constant observation of predicates that accompany their subjects as attributes or modes;
3. axioms and postulates, that is, statements from whose content it is possible to immediately conclude whether or not the subject is contradicted or not contradicted by the predicate;
4. previously proved theses.

Wolff picked up (although he also was unable to realize it in any substantial form) the Leibnizian concept of a "universal characteristic" and curiously defined this "characteristic" as follows: "The art that considers the signs convenient for discovery, the methods for combining them, and a definite law for changing these combinations is called the art of the combinatorial characteristic. It is also called by Leibniz the general theory of forms (*speciosa generalis*). Part of this combinatorial characteristic is the field of algebra that has been felicitously (under the influence of Leibniz) called the art of the combinatorial characteristic of quantities. The ideas of this science were already known to Leibniz in a relatively developed form; he successfully penetrated deeply into the nature of algebra" (Reference 195, Section 294).

Wolff drew parallels between (1) arithmetic calculation with numbers (*numerorum in Arithmetica*), (2) algebraic transformation of magnitudes (*magnitudinum in Algebra*), and (3) syllogistic conclusions (*syllogismorum in Logica*), although he saw certain difficulties associated with the application of the procedures of (1) and (2) in logic. He was led to conclude that the objects of logic have first, as it were, to be converted before they are amenable to the methods of mathematical calculation. In essence, this is no more than a statement of the difficulties with which Leibniz met in his logical calculuses.

Wolff actually associated the term "characteristic" (*Characteristica*) with the notion of a logical algorithm, although his discussion of this problem was, to a considerable extent, complicated by the fact that it appeared in the context of the more general problem of creating a universal philosophical language.

Wolff's students Bilfinger (1693–1750), Baumgarten (1714–1762), Baumeister (1709–1785), and Meyer (1718–1777) continued along the paths of theoretical logic discovered by Leibniz and their master but also applied logical ideas to humanitarian and pedagogical ends. These thinkers did their best to bury any idea of making logic mathematical. It was from their scanty and confused logical compendia (such as Meyer's *Vernunftlehre*, 1762) that Kant was later spoon-fed formal logic.

The only work that rose above this level was Reusch's *System of Logic* (1760), which was apparently a test of the influence of Hansch's *Art of Invention* (*De arte inveniendi*, 1727).

Sporadic attempts at introducing logical operations analogous to arithmetic operations were made by the philosopher Joachim Georg Darjes (1714–1791) of Jena. He made his first attempt in *Introductio in artem inveniendi seu logicam theoretico–practicam*, published in Jena in 1747, and a second in a "modest" publication titled *Path to Truth . . .*, published in Frankfurt in 1744. Methodologically, Darjes tended toward the ideas of the German Age of Reason, while his logic tended to follow that of Pierre Gassendi.

As a sample of Darjes' logical symbolism, we present his notation for a universally negative proposition, which he wrote in the form

$$+S - P.$$

This means that if the subject S is accepted, the predicate P must be eliminated. This representation is close in meaning to the implication $S \rightarrow \bar{P}$.

Darjes' algorithms lacked the necessary degree of generality, and it was only with difficulty that they were applicable even to formalization of syllogistics. The historical survey appended by Darjes to Reference 68 was composed of texts by Gassendi.

Leibniz' logical ideas, however, were not completely dead, for the German mathematician J. A. Segner (1704–1777) returned to them again, along with his countryman, the logician and philosopher Gottfried Ploucquet (1716–1790).

Johann Andreas von Segner, the author of *Cursus mathematici*, which was well known in its time, was Professor of Mathematics at Göttingen and then at Halle. While literal calculus was only a generalization of numerical arithmetic in the mathematical works of Newton, for Segner it was distinct from mathematics. Segner was attracted to a scientifically systematic approach to algebraic calculus (a path that had already been tried by Wolff), and he attempted to establish individual rules for constructing rigorous proofs. However, the arithmetic during his era had not yet been provided with any rigorous theoretical basis, so his attempts were naturally doomed to failure.

Segner devoted much attention to developing methods for graphical and numerical solution of equations. Jakob Bernoulli had attempted to construct an equation of the form

$$a = bx + cx^2 + dx^3 + \cdots + px^n$$

directly by means of the points of intersection of the line $y = a$ and the curve

$$y = bx + cx^2 + \cdots + px^n.$$

Segner simplified his method in *Nov. Comm. Ac. Petr.* (1758–1759). Beginning by finding the points of intersection of the line

$$y = -a + bx + cx^2 + \cdots + px^n$$

with the x-axis, he found a relatively simple construction of such a curve. Segner is also known for development of the theory of parabolic curves (Reference 286, p. 60). In *Nov. Comm. Petr.* (1758–1759), he proposed a special recurrent device for solution of problems dealing with the complete characteristic — the method of the diagonal partitioning of n-cornered figures into triangles." He discussed spherical trigonometry analytically (Reference 286).

The logical results of Segner's work are presented in his Latin treatise *Specimen logicae universaliter demonstratae* (Jena, 1740). He was one of the earliest authors to make broad use of a system of symbolic notation for logic. For example, he used the notation $-X$ to denote the negation of the concept X. His use of this symbolism did not require that the symbol $-X$ always be used for notation of so-called negative ideas of general logic — it could also be used in expressions containing the particle "not." Thus, if Z denotes the class of nonpolygons, the class of polygons can be denoted by $-Z$ (Reference 200, p. 71).

At the base of Segner's calculus lay the thesis that an individual letter must replace an entire word (concept). According to Segner, the expression $A < B$ indicates that the concept A is included in the concept B. The universally negative proposition "no S is P" of syllogistics takes the form $S < -P$; here S is included in "not P." He did not limit himself to formalization of the classical syllogistic conclusions, and his work contains examples of conclusions beyond this narrow framework. For example, from the premises $A < B$ and $C < D$, Segner concluded $AC < BD$, where the omission of a symbol between the letters must be taken to indicate the operation of intersection of classes.

This Göttingen mathematician was familiar with the law of idempotence $X \cdot X = X$. This is evident from his comment "It is clear that an idea added to itself cannot become the subject of a new definition; a line has extension, a curve has extension, and so a curved line has extension; here the predicate, added to itself, does not change" (Reference 200, p. 148).

Segner spoke of an addition of properties that corresponds to the operation of multiplication of classes. Setting $C = A$ in the syllogism

$$((A < B)(C < D)) < (AC < BD),$$

he obtained

$$((A < B)(A < D)) < (AA < BD).$$

But $AA = A$, from which it follows that

$$((A < B)(A < D)) < (A < BD).$$

Thus, Segner actually discovered a number of theorems of modern propositional calculus.

Segner did not distinguish between symbols for relations and symbols for operations. In the expression

$$((A < B) \cdot (C < D)) < (AC < BD),$$

the symbol $\cdot$ on the left actually plays the role of a conjunction (but not of the operation of the intersection of classes), while the third symbol $<$ from the left is used to denote derivation (the relation of implication). In modern notation, Segner's proposition would take the form

$$((A < B) \,\&\, (C < D)) \rightarrow (AC < BD).$$

Segner considered a series of similar statements, the majority of which, however, were known to Leibniz. For literature concerning Segner, see References 201–203.

In his logical theory, the Leibnizian Gottfried Ploucquet used the idea of a "universal characteristic" as a starting point, although he took it as a problem of constructing a sufficiently strong logical calculus in which, in particular, it would also be possible to derive meaningful facts of traditional logic. His conception was associated with representation of various elements of propositions by geometric figures and algebraic formulas and with application of a geometric language to analysis of propositions (including syllogisms). According to Ploucquet, the "universal characteristic" should not be treated as requiring construction of a universal metaphysical language but should be considered only with respect to its role in logic.

Gottfried Ploucquet was born August 25, 1716 of a French family that immigrated to Germany for religious reasons (his family was Calvinist). In Germany, Gottfried's father became the manager of a large hotel. His mother, Elizabeth Eleanor, was German, and the native language of the son was German.

Gottfried's parents sent him to the ducal high school at Stuttgart, where his extraordinary abilities immediately attracted the attention of his teachers. In 1732 the sixteen-year-old enrolled at Tübingen University, where philosophical training was treated only as a preliminary to theological speculation. Of Ploucquet's teachers at the university, the most influential was Israel Gottlieb Canz, who introduced Ploucquet to the works of Liebniz and Wolff. Ploucquet studied their mathematical works initially, then their philosophical works.

In 1734 Ploucquet passed his Master's examination and proceeded to the study of theology. His student years, especially those devoted to theological affairs, fell into the period during which the church and theologians were dominated by the spirit of Pietism. In his world view, Ploucquet followed this ideological direction, which clearly expressed the interests of the German burghers in their struggle against traditional feudal institutions. This circumstance had no little effect on the ideological cast of Ploucquet, a fact noted by his biographer, Karl Aner (Reference 204, p. 8). In comparison with the orthodox Pietists of Halle (who, at the time, were persecuting Wolff), the Wurtemburg Pietists were distinguished by their relative tolerance and high degree of democracy.

Fortunately for science, Ploucquet minimized the time he spent on theological problems. He was increasingly attracted to subtle theoretical problems, the most important among them being his relationship to the methodological heritage of the leader of German rationalism, Leibniz.

In 1749 his publication *Primaria monodologiae capita* (1748) won him a membership in the Berlin Academy of Sciences. In this work he recommends Leibniz' monadology as a satisfactory philosophical theory.

Beginning in 1750, Ploucquet's activities were associated with the University of Tübingen. The scientific interests of the new academician were concentrated on logic and philosophy, while his theological interests cooled completely.

In 1763 Ploucquet, by now Professor of Logic and Philosophy, became rector of the University of Tübingen. He set about reconstructing Leibniz' ideas of a logical calculus and entered upon a lively correspondence on logical problems with Lambert and his student Holland. In his views of psychology Ploucquet leaned toward the rationalistic school, while in biology he was attracted to Leibnizian preformism. At the same time, he gained a certain notoriety as a commentator on classical philosophical texts: He published outstanding commentaries on Thales, Anaxagoras, Democritus, Pyrrho, and Sextus Empiricus. He died on September 23, 1790 in Tübingen.

In philosophy, Ploucquet was a typical thinker of the Leibniz–Wolff school. For example, his definition of the categories of necessity and contingency follows that of Wolff, as we see from his definition: "necessity is that which, if contradicted, contradicts the principle of identity; contingency is that for which the preceding characteristic is not true" (Reference 205, Section 157). From Wolff he drew the idea that certain forces are concealed in matter. In a discussion of the relationship between the finite and the infinite, he stated that the "finite, added to the finite, does not yield the infinite" (see Reference 205, Section 77) and thus leaves the reader in ignorance about the origin of the category of infinity.

Of the logical works of Ploucquet, we should note the following: *Methodus am demonstrandi directe omnes syllogismorum species quam vitia formae detegendi ope unius regulae* (Tübingen, 1763); *Untersuchung und Abänderung der logikalischen Konstruktionen des Herrn Professors Lambert* (Tübingen, 1765); *Sammlung der Schriften, welche den logischen Kalkül des Herrn Prof. Ploucquets betreffen, mit neuen Zusätzen hrsg. v. Fr. Augustus Böck* (Frankfort and Leipzig, 1766); *Antwort auf die von Herrn Professor Lambert gemachten Erinnerungen und Diesseitiger Beschluss der logikalischen Rechnungsstreitigkeiten* (1766); *Institutiones philosophiae thereticae* (Tübingen, 1766). This work contains material on a logical calculus. *Elementa philosophiae contemplativae* (Stuttgart, 1788). This work is partially concerned with problems of logical formalisms. *Expositiones philosophicae theoreticae* (Stuttgart, 1782). This work contains an essay of a logical nature. It also presents a final form of attempts to rest logic on a basis of geometric principles by associating topological elements with concepts. For literature on Ploucquet, see References 204, 206–217. In the first edition of *Principia de substantis et phenomenis metaphysica* (1763), the logical texts are absent, and logic itself is treated only very generally. For

Ploucquet, logic was split into two studies: the first considers the foundations and principles of arguments, while the second considers the foundations and principles of method. In his second publication, *Principals* . . . (1764), Ploucquet had already formulated, in particular, a theory of sentences. A sentence establishing identity between a subject and a predicate is affirmative; a sentence establishing a difference between a subject and a predicate is negative. This concept of logic is called "the theory of identity of concepts associated in statements." Ploucquet was not the first to develop it, and prior to him, it was discussed in the works of G. G. Titius, *Ars cogitandi* (1701), pt. 7, sec. 53 ff., and A. Rüdigeri, *De sensu veri et falsi* (1722), pp. 231 ff. (The first edition of this work appeared in 1709. There was also a third edition in 1741.) See also A. Arnauld and P. Nicole's *La logique ou l'art de penser, contenant autre les règles communes* (Paris, 1644). Ploucquet himself sought the origins of his theory in Leibniz' works and devoted his article *Anmerkungen über Leibnizens* "*Difficultates quedam Logicae*" to this topic (see Reference 218, p. 260).

Leibniz sometimes wrote the universal affirmative proposition "all A are B" in the form "all AB are A." To provide his assertion of this with some basis, Ploucquet noted that "since Leibniz had noted precisely this fact (that is, the possibility of expressing a universally affirmative proposition in the above manner), he was forced to express the identity of the subject and predicate in affirmative propositions" (Reference 218, p. 260). A predicate, Ploucquet continued, cannot differ from its subject, for otherwise the subject would not be the subject, a condition which is impossible. One must therefore posit the theory of identity of logical elements as the basis of logic.

While affirmative statements express identity, negative statements express the difference between the concepts combined in them. Ploucquet wrote universal negative propositions in the form $S > P$ (in words, S and P are different, that is, no S is P). The operation $>$ (a symbol introduced by Ploucquet himself) has the property of symmetry. Thus, if $S > P$ is true, so is $P > S$. It is easy to see that Ploucquet's binary symmetric operation $>$, which denotes negation of the conjunction of S and P (the formulas

$S \rightarrow \bar{P}$ and $\overline{S \& P}$ are equivalent!), is, in its formal properties, identical to the Sheffer stroke.

According to Ploucquet, a complete description of all logical relations requires only two functors — the identity functor and the inconsistency functor. Logic can conclude only that the objects it considers are identical or different. To obtain more information, it is necessary to leave the frigid realm of logic for the tropical fertility of concrete scientific disciplines. Of course, this doctrine did not agree with the tastes of Hegel and the Hegelians. Ploucquet's logical calculus was not only complex, but, if we may say so, multileveled. Among its many levels, we find represented the logic of classes, the propositional calculus, the beginnings of the theory of relations, and generalized syllogistic constructions.

The alphabet of Ploucquet's calculus contains capital letters $A, B, C, D, \ldots$, and lower-case letters $a, b, c, d, \ldots$, which, respectively, are used to denote distributed and undistributed terms.

The symbols used for binary relations are the symbol of identity (*est*) and the symbol for distinction or incompatability ($>$). The relation *est* is sometimes denoted by omission of any symbol between a capital letter and a small letter following it (for example Ac) and sometimes by an expression of the form $(A - c)$.

As we have already noted above, the relation $>$ is formally identical to the Sheffer stroke (in the sense of negation of conjunctions). It is symmetric but not transitive. "The expression $abc > de$ means that d, which is combined with e, is negated with respect to a, which is associated with b and c" (Reference 207, p. 75).

The operations used in the calculus are as follows:

1. The symbol $+$ is used for conjunction (in the logic of propositions) and the operation of intersection of classes (in the logic of classes).
2. The multiplication sign, placed between small letters, or the absence of such a sign between lower-case letters, denotes the operation of inclusive disjunction (for propositions) and the operation of union of classes without elimination of common members.

Concerning this operation, Ploucquet remarked, "for me, the expression mb does not represent some form of multiplication, but must be taken to mean some association of concepts" (Reference 218, p. 254). In using the word "representation," Ploucquet had in mind a definite property corresponding to some class (volume).

3. The unary operation of negation is indicated by a bar over a capital or small letter. The expression $\bar{P}$ denotes "all not P," while the expression $\bar{p}$ denotes "some not P." Ploucquet did not use the representation of the negation of a concept X by means of the expression $X > X$ or, equivalently, $1 > X$, where 1 denotes the identically true constant.* The process of constructing conclusions is described by Ploucquet in terms of substitution of identicals for identicals and distinction of nonidenticals. We should note the following series of rules of inference that he used:

1. "If a first proposition† is negated by some second proposition which, in turn, is negated by some third proposition, where the third and first propositions possess a certain common part, then it follows that the second proposition negates the disjunction of the first and third propositions. For example, from the premise $ab > c > deb$ it follows that $abde > c$ (Reference 207, p. 77).

Taking the absence of symbols between the lower-case letters as a sign for inclusive disjunction and $>$ as the Sheffer stroke $|$, we can show that, in fact, from the premises $(a \vee b) \mid c$ and $(d \vee e \vee b) \mid c$ we can derive the conclusion $(a \vee b \vee d \vee e) \mid c$. We obtain

$$(a \vee b) \vee (d \vee e \vee b) \mid c,$$

that is,

$$(a \vee b \vee d \vee e \vee b) \mid c \quad \text{or} \quad (a \vee b \vee d \vee e) \mid c,$$

which is what we wished to prove. Everywhere below, we have replaced Ploucquet's sign $>$ with the equivalent and more familiar sign $|$.

2. According to Ploucquet's second rule, if the relation

* We should note a certain evolution in Ploucquet's use of symbols. Thus, for example, in his early works he used the expression $OM - qP$ to represent "Every M is some P" (*Omnis M est quaedam P*); later he wrote the same proposition as $M - p$ and, finally, simply as Mp, denoting, in essence, the equality $M = p$.

† Ploucquet uses the expression "subject" in the actual text, but since the text concerns the logic of propositions, he certainly has statements in mind.

$(a \vee b \vee c) \mid (d \vee e)$, is true, then the relations: $a \mid d$; $a \mid e$; $b \mid d$; $b \mid e$; $c \mid d$; $c \mid e$ are true. The expression $(a \vee b \vee c) \mid (d \vee e)$ is equivalent to the formula $(a \vee b \vee c) \rightarrow \overline{(d \vee e)}$, where $\rightarrow$ denotes implication. Hence we have $(a \vee b \vee c) \rightarrow \bar{d} \;\&\; \bar{e}$, which is equivalent to the system of premises $a \rightarrow (\bar{d} \;\&\; \bar{e})$; $b \rightarrow (\bar{d} \;\&\; \bar{e})$; $c \rightarrow (\bar{d} \;\&\; \bar{e})$. But from $a \rightarrow \bar{d} \;\&\; \bar{e}$ it follows that $a \rightarrow \bar{d}$ and also $a \rightarrow \bar{e}$. Similarly, $b \rightarrow \bar{d} \;\&\; \bar{e}$ implies $b \rightarrow \bar{d}$ and $b \rightarrow \bar{e}$. Finally, from $c \rightarrow \bar{d} \;\&\; \bar{e}$ we obtain $c \rightarrow \bar{d}$ and $c \rightarrow \bar{e}$. This implies the truth of the desired relations $a \mid d$; $a \mid e$; $b \mid d$; $b \mid e$; $c \mid d$ and $c \mid e$. For the formulation of this rule see Reference 207, pp. 77–78.

3. The third rule refers to conjunctive addition of two "associations." It asserts that $(a \vee b) \;\&\; (c \vee d)$ (in Ploucquet's notation, $ab + cd$) implies the conclusion $(a \vee b) \vee (c \vee d)$ or $a \vee b \vee c \vee d$. Of course, this is true. However, even Geulincx knew that $A \;\&\; B$ implies $A \vee B$.

4. The fourth rule is as follows: Given the premise

$$(A = b) \;\&\; (C \mid D) \;\&\; (A \mid c),$$

we can derive from it the conclusion $(A = b = C) \mid D$. Proof: The expression $(A = b) \;\&\; (C \mid D) \;\&\; (A = c)$ is represented in the form $(A = b = c) \;\&\; (C \mid D)$, but since c is subordinate to C (the quantifier "some" is subordinate to the quantifier "all"), $(A = b = c) \mid D$, which was to be proved.

Thus, the first three rules concern the logic of propositions, and the fourth concerns the calculus of classes with the identity relation. In the latter case, $X \mid Y$ should be understood as an abbreviation for the expression: "X is contained in the complement of Y"; that is, the volume of class X lies outside of the volume of class Y.

By using Ploucquet's first three rules, we can derive a number of relations with the functor $\mid$ in the propositional calculus. For example, setting $e = c = 0$, where 0 denotes the identically false constant, we can obtain from Ploucquet's second rule the following:

5. $((a \vee b) \mid c) \rightarrow ((c \mid a) \;\&\; (c \mid b))$;

the dual of this takes the form

6. $((a \;\&\; b) \mid c) \rightarrow ((c \mid a) \vee (c \mid b))$.

Ploucquet did not explicitly state the laws $X \& X = X$ and $X \vee X = X$, but they were actually known to him, as is indicated by his use of them in his calculus. In one place, having obtained the equation $nA + nA = m + b$, he continued: "since the expression $nA + nA$ represents a repetition of one and the same symbol (*Zeichenwiederholung*) . . ., it therefore follows that $nA = m + b$." (Reference 115, p. 254). He discussed in detail the Leibnizian concept of the inclusion relation, that is, the inclusion of the property X in some Y, as indicated by the equality $X \vee Y = Y$. Ploucquet also indicated that this relation was rarely used, even by Lambert (Reference 218, pp. 262 and 212).

Ploucquet attempted to introduce a number of corrections to traditional syllogistic theory, in particular to that rule which states it is impossible to obtain a necessary conclusion from two negative premises. Thus, he considered it possible, given the premises "All A are not-B" and "All A are not-C," to obtain the conclusion: "Some not-C are not-B." His remark has meaning, however, only if we treat the statement "X is not-Y" as negative. In this sense, Ploucquet has implicitly noted that the concept of the complement of a class is absent in Aristotelian logic; this concept was rigorously defined only later by De Morgan in the nineteenth century.

Following Lambert, Ploucquet laid the foundations for a generalized syllogistic theory, placing the quantifiers "all" and "some" before not only the subject but also the predicate of a statement. Ploucquet demonstrated (Reference 218, p. 212) that the operation of quantification of the predicate was used by Lambert.

Beginning in 1758, Ploucquet also worked on the idea of a topological interpretation of logical relations, placing geometric elements in correspondence with logical terms.

Ploucquet's logical calculus was adopted in its entirety by Heinrich Wilhelm Glemm, a professor and lawyer from Stuttgart, as well as by the German mathematician G. J. von Holland and the writer Thomas Abbt. However, Lambert and Segner were both critical of some aspects of Ploucquet's work. Lambert used the exclusive "or"; Ploucquet used inclusive disjunction. Their disagreement recalls the later (nineteenth century) divergence in the treatment of disjoint propositions by George Boole and W. S. Jevons.

The polemics between Lambert and Ploucquet also related to the treatment of universally affirmative propositions. Whereas Lambert considered the predicate to be the generating concept for the subject of a (universally affirmative) statement, Ploucquet, on the contrary, asserted that the idea of such a statement reduces to the following: Every A belongs to a certain B, where B is not a generating concept, but obtains its meaning from the subject. Here Ploucquet's view is closer to that of Leibniz. In other words, the thesis "A *est* B" appears in the works of Ploucquet in the form $A \vee B = B$, where the left-hand side of the equality is the subject and the right-hand side is the predicate. Lambert was inclined to interpret the same statement in the form $A = v \cdot B$, where A is the subject and $v \cdot B$ is the predicate. From our point of view, the difference between these two treatments is of only a linguistic nature and can hardly be considered an adequate basis for discussion.

Ploucquet's logical calculus, by its very nature, is neither a universal language in the Leibnizian sense nor a calculus of concepts in the sense of the logic of classes. It can be considered only as the beginnings of a unified calculus of statements and one-term predicates; one of the important fragments of Ploucquet's calculus is his introduction of a symbol equivalent to the Sheffer stroke in propositional logic. Ploucquet influenced the logic of J. Lambert (who in turn influenced Ploucquet) and G. von Holland. The nineteenth-century German logician C. August Semler was also well informed as to the beginnings of a calculus of relations in Ploucquet's work (see Semler's *Versuch über die combinatorische Methode*, Dresden, 1811).

It is interesting that Ploucquet, unlike Leibniz and the brothers Bernoulli, does not consider inverse operations in logic (he does not mention the operations of logical division and logical subtraction). Perhaps he saw no way of resolving the difficulties which arise in such a consideration. By eliminating such problematic questions and by actually introducing the Sheffer stroke, Ploucquet took a definitive step away from the logic of Leibniz and toward the foundations of contemporary logical formalisms.

The philosophical contemporaries of Ploucquet met his logical investigations without enthusiasm. The opinion of Hegel (Reference 215) is indicative of this. Concerning Ploucquet's declaration that even people not versed in logic could learn it quickly by using his logical calculus, Hegel, up in arms against the impudence of such an idea, said: "Such an idea represents, of course (?) [question mark is mine — N. S.], the worst that can be said about any invention for elucidating logical science" (Reference 215, p. 133). Since even ordinary formal logic "didn't suit" Hegel, it is not surprising that its algebraicized version was not to his taste either. Hegel's evaluation of the logical calculus of his countryman Ploucquet is interesting: "This calculus is based on a process of taking a statement and abstracting from the distinction in relations, that is, the distinction between uniqueness, singularity, and universality; it fixes an abstract identity of the subject and predicate, on the strength of which a mathematical equality is established between them — this transforms the process of syllogism into a completely empty and tautological structure of propositions" (Reference 215, pp. 132–133). One can only imagine what epithets Hegel would have bestowed on contemporary mathematical logic! As far as Hegel himself is concerned, he was able to adapt only the forms of school logic to the rather unclear language of his own philosophical system.

Of the later philosophers, B. Erdmann (Reference 216) was the first to give serious attention to the logical meaning of the scientific investigations of Ploucquet.

2. The Development of the Propositional Calculus and Logic of Relations in the Work of the Eighteenth-Century Logician Johann Lambert

Of the many creative followers of the logic of G. W. Leibniz, Johann Heinrich Lambert, without a doubt, occupies the most prominent place. In the logical works of this well-known German philosopher, mathematician, physicist, and astronomer of the nineteenth century, we already find a developed system of logical calculus, many of whose traits anticipate some of the characteristics of much later logical–mathematical formalisms.

Johann Heinrich Lambert was born in 1728 in the small town of Muhlhausen in Upper Alsace, which was at that time part of the Swiss Confederation. He was French by ancestry. Lambert's family lived under rather difficult material circumstances. The sixteen-year-old Johann had to take a job as a school clerk, which he later left for a position as secretary to Professor Iselin of Basel. This gave the inquisitive youth an opportunity to begin work on his own systematic education.

Reflecting on ethical problems (in the spirit of rationalism), Lambert came to the conclusion that the perfection of morals was impossible until such time as they were subject to the impartial judgment of the intellect. He became an avid reader of philosophical literature, studying Wolff (*On the Powers of Human Reason*), Malebranche (*On the Criteria of Truth*), and Locke (*Essay Concerning Human Understanding*). At the same time, he studied mathematics and natural sciences. "Mathematical sciences, in particular algebra and mechanics," notes Lambert in one of his letters of 1750, "gave me a clear and convincing example for verification of rules being investigated [in philosophy — N.S.]."

In 1748, on the recommendation of Professor Iselin, Lambert became a tutor in Count Salis' family in Chur. There he continued his study of mathematical and natural sciences.

In 1756, Lambert went on a journey with his charges. The itinerary included Göttingen, Utrecht, Leyden, Amsterdam, The Hague, Paris, Turin, and Milan, with a final return to Chur. After a visit to his native town, in 1759, Lambert went to Augsburg, where his famous treatise in physics, *Photometry* (1760) was published. This book contains Lambert's formulation of the law establishing that the intensity of the light emitted from several luminous surfaces depends on the direction in which the measurement is made.

Upon arriving in Munich, Lambert became an acting member of the local Academy of Sciences. In his *Cosmological Letters* (1761), he appears before us as a typical thinker of the Leibniz–Wolff school. In his scientific views, Lambert leaned heavily on the results of Newton; in his cosmogonical views he was close to Kant. Lambert investigated the orbits of comets and laid the foundations for a theory of double stars.

After losing his place in the Munich Academy, Lambert was chosen an acting member of the Academy of Sciences in Berlin (in 1765).

In 1766, Lambert, basing his work on some of Euler's results, first proved that the number π can be expressed as an infinite non-repeating decimal (later his proof was improved somewhat by the French mathematician Legendre). Lambert was also the first to make systematic use of hyperbolic functions.

As a philosopher, Lambert was of a very pronounced synthetic turn of mind. His philosophical views were based on materialist premises. In his correspondence with Kant, he insisted on the necessity of supplementing the rationalism of Wolff with the empiricism of Locke. In *Neves Organon* (Leipzig, 1764) and *Architectonics* (Riga, 1771), Lambert attempted a systematic structuring of philosophy. Philosophy, using Lambert's own terminology, could be subdivided into the following four parts.

1. *Dianoetics*: the science of the laws of human understanding, close to psychology. It concerns actual thought processes; that is, the latter are studied from the point of view of their continuous flow in reality.

2. *Alephiology*: the science dealing with the means of revealing elementary concepts, the laws of construction of complex concepts from elementary ones, and the necessary notation for them both. In this respect, Lambert can be considered one of the founders of contemporary semiotics as a universal theory of notation (outside of the relationship of the latter to the objects being denoted).

3. *Semiotics*: the science of the relationship between things and ideas (this correlates with contemporary logical semantics as a science dealing with the relationship between notation and what is being denoted, between formal systems and their interpretations, etc.).

4. *Phenomenology*: the study of "being" (languishing, from the contemporary point of view, because of its metaphysical boundedness, that is, because of the well-known disparity between "being" and "essence"). This would be similar to the later subjectivist phenomenology of Husserl and Jaspers.

In logic, Lambert, following Leibniz, became one of the pioneers in logicomathematical conceptions, anticipating many of the later results of George Boole. In particular, Lambert was apparently aware of the procedure for decomposing a logical expression into elementary components; he also carried out a logical analysis of the strict exclusive union "or" and studied the basic inverse operations of the calculus of classes. A certain defect in Lambert's calculus consists in the fact that sometimes his extrapolations of mathematical (more precisely, algebraic) conclusions to the realm of formal logic are not sufficiently well founded. The logic developed by Lambert included notation for classes and symbols for the various relations. Lambert also gave an original construction for a generalized syllogistic theory and, by quantifying the predicate, anticipated many features of subsequent studies in deduction by English logicians of the nineteenth century.

From 1763 until his death in 1777, Lambert lived in Berlin, supported by a special academic pension.

Of the philosophical and logical works of Lambert, we make note of the following: *Neues Organon*, 2 vols. (Leipzig, 1764). Here Lambert presents a notation system sufficient for formalizing syllogistics with its four figures. The first volume contains a theory of geometric (linear) interpretation of logical forms. The beginnings of a theory of chains of syllogisms is presented.

Neue Zeitungen von gelehrten Sachen (Leipzig, 1765). Here Lambert discusses the logical calculus of Ploucquet. "De universliori calculli idea, disquisitio una cum adnexo specimine," *Nova acta eruditorum* (Leipzig, 1765); "In algebram philosophicam Cl. Richeri breves adnotationis," *Nova acta eruditorum* (Leipzig, 1767); *Brief von Verbesserung der Logik durch mathematische Ausdrücke*, *Göttingische Anzeigen von gelehrten Sachen* (5 March, 1764); *Anlage zur Architectonic oder Theorie des Einfachen und des Ersten in der philosophischen und mathematischen Eřkenntniss*, vols. 1–2 (Riga, 1771). From the point of view of logic, the second volume is much more interesting than the first. In the first volume, modal deductions are analyzed; in the second, a theory of identity and basic logical operations is presented. *Logische und philosophische Abhandlungen, hrsg. von J. Bernoulli*, 2 vols. (Dessau, 1782–1787) — a Berlin edition

(1792) also exists. These volumes contain most of Lambert's logical discoveries. For literature on the logical and philosophical views of Lambert, see References 219–237.

J. H. Lambert's logical calculus is very difficult to present and interpret, because the language of this calculus and the methods employed can easily strike the contemporary reader as artificial and unfounded. Thus, along with logical variables and functors for copulas, Lambert introduced numerical coefficients, symbols for fractions, etc., which appear to be uninterpretable with respect to logic. Nevertheless, Lambert's results stand much closer to the present-day form of symbolic logic than does, for example, the calculus of G. W. Leibniz.

Many authors (see, for example, Reference 226, p. XXXIII) correctly note many similar traits in Lambert's calculus and in that of the English logician George Boole. And, in fact, as we will point out later, the nature of the difficulties in these two logics is also analogous.

Lambert distinguished (Reference 239) two types of symbols: (1) for logical classes (or concepts) (*Zeichen der Begriffe*) and (2) for denoting logical operations (*Verbindungskunst der Zeichen*). Concepts, in turn, were divided into three groups: (a) known concepts (denoted by the initial letters of the Latin alphabet: $a, b, c, d, \ldots$); (b) undetermined concepts (denoted by the middle letters of the same alphabet: $n, m, l, \ldots$); and, finally, (c) unknown concepts (denoted by the last letters of the Latin alphabet: $x, y, z, \ldots$) (Reference 238, vol. 1, sec. 9, p. 5).

The introduction of undetermined concepts was obviously necessitated by Lambert's consideration not only of direct logical operations but also of inverse operations; the latter are characterized by a certain indeterminateness. The discrimination between known and unknown concepts was dictated, no doubt, by the analogous subdivision of variables in algebra.

At the base of Lambert's calculus, we find the following four operations: (1) combination (*Zusammensetzung*), (2) isolation (*Isolierung*), (3) definition (*Bestimmung*) and (4) abstraction (*Absonderung*). In their formal properties, these can be considered analogous (although not identical) to the arithmetic operations of addition, subtraction, multiplication, and division, respectively (Reference 238, vol. 1, p. 150). Lambert asserts that combination and isolation are inverse operations and that definition and abstraction are inverse operations—just like addition and subtraction, and multiplication and division, in arithmetic algebra.

Thus, we are dealing with two direct (combination and definition) and two inverse (isolation and abstraction) operations. To avoid confusion, we will call combination *logical addition* (Lambert denotes it by the sign $+$); definition *logical multiplication* (Lambert denotes it by either the sign $\times$ or the absence of any sign between letters). Analogously, we will call the operation isolation *logical subtraction* (denoted by $a - b$), and the operation abstraction *logical division* (denoted by $a \div b$).

The expression $a + b$ represented for Lambert strict disjunction (in propositional logic) and symmetric difference (the union of classes with exclusion of their common part) in the calculus of classes. Formulas of the form $a + b$ represented "disjunctive propositions, not requiring geometric illustration because they do not express anything positive" (Reference 239, sec. 190). In other words, Lambert considered, in disjunction, only the case of mutually exclusive alternatives.

The operation $-$ in the expression $a - b$ was understood to be an abbreviation for the combination of words "those a which are different from b." Lambert noted that $a - b$ is logically interpretable only under the condition that b does not exceed a. "Our metaphysical symbolic art," he writes, "differs markedly from a mathematical calculus . . ., in that in our system the subtrahend cannot exceed the minuend." (Reference 240, vol. 2, p. 62.) In logic there is no place for negative quantities in the algebraic sense. In another chapter, the German logician wrote $a - b$ in the form a/b (literally: a and not-b). Lambert was aware of the procedure for decomposing logical expressions into their constituents (units). Thus, for example, he writes

$$a = ax + a/x, \tag{1}$$

where the expression a/x denotes the class of subjects a not having the property x. Therefore, John Venn correctly noted that Lambert "decomposes . . . logical expressions exactly as does George Boole, although he does not derive for them general formulas of decomposition" (Reference 226, p. XXXIII).

The expression $a \times b$ was construed by Lambert as conjunction (in propositional logic) and as intersection of volumes (in the calculus of classes). Sometimes, instead of $a \times b$ he wrote $a \cdot b$ or simply ab.

The operation of logical division $\div$ he described as an operation inverse with respect to the operation $\times$. In other words, the expression $x = a \div b$ was actually understood by Lambert as equivalent to the equation $b \cdot x = a$. Just as in arithmetic the expression $\frac{a}{b}$ can be construed as denoting some number which, when multiplied by b, yields a, similarly Lambert's expression $a \div b$ denotes that class of subjects which, when intersected with b, forms the class a.

We will not now analyze in detail the difficulties connected with the treatment of the operation of logical division; we only note, first of all, that the operation $a \div b$ in the calculus of classes has meaning only provided that a does not exceed b and, secondly, that this operation is not unique, as is well known. This lack of uniqueness can be illustrated by a specific example. Let $a = b = o$ (here o represents the empty class), then the expression $o \div o$ is undefined in the sense that it denotes *any* class k (any class satisfies the equation $k \cdot o = o$). Lambert used the notation $\frac{a}{b}$ as well as the notation $a \div b$ to indicate logical division.

In addition to the symbols for operations ($+$, $-$, $\div$, and $\times$), Lambert used the following additional symbols for relations: (1) equality ($=$) as an abbreviation for the copula "is" in the sense of identity; (2) inequality ($- =$) as a symbolization of the expression "is not" (the negation in the statement is relative to the copula). Lambert notes in this connection: "I will express the copula 'is' by using the equality sign, and the copula 'is not' by using the arithmetic sign of negation . . . The proposition 'not a single A is M' will be expressed as: $A = -M$" (Reference 241, vol. 1, p. 396).

We should also note a number of tautologies established by

Lambert which relate not only to the calculus of classes but also to propositional logic (below + denotes strict (exclusive) disjunction):

$$a + b = a/b + b/a + ab + ab, \quad \text{(Reference 238, vol. 1, p. 11)}; \tag{2}$$

$$a + x/a = x + a/x; \tag{3}$$

$$a + a/b = ab. \tag{4}$$

On the right-hand side of (2) the last two terms could be eliminated since for strict disjunction the relationship $X + X = 0$ holds. However, we do not find this law anywhere in Lambert's work. Furthermore, the German logician was sometimes inclined to write the sum $x + x$ in the form $2x$, although it would seem that even coefficients of x should not appear at all in his work. Surprising as it may seem, this error does not cause troublesome difficulties in Lambert's calculus. He considered it possible, from the truth of the expression $2y = x + y$, to conclude the truth of the equality $y = x$; that is, he somehow subtracted y from both sides of the equation. Jorgen Jorgensen (Reference 225, pp. 82–83) sternly reproaches Lambert for this, observing that such a procedure is logically inadmissible and that Lambert here illegitimately transferred to logic the rules of arithmetic.

However, let us consider what actually takes place during such a transformation by Lambert. Since $2y$ is in fact equal to 0, then, by concluding from the truth of $0 = x + y$ the truth $y = x$, we are indeed only stating the fact that, if the exclusive disjunction of x and y is false (that is, empty), then the assertion of the equivalence of x and y must be true. No one would quarrel with the logical validity of this conclusion. Thus, the use of coefficients in Lambert's logic can be eliminated. Even if they remain as terms in disjunctions, such coefficients don't spoil anything (false terms can be excluded from a disjunction without changing the meaning of the disjunction as a whole).

Lambert anticipated Boole's law of idempotence $x \times x = x$; however, Lambert's version did not have universal applicability in his calculus since the calculus developed by him included, in addition to symbols for classes, symbols for relations (see Reference 234, pp. 96–97).

As a good example of the problematics Lambert was able to work out in his calculus, let us consider the following logical problem which he poses (and solves):

If that part of X which is not-A coincides with B, and that part of A which is different from X is identical with C, how can we express X in terms of only A, B, and C? (Reference 238, vol. 1, p. 14).

Lambert sought a solution in the form "$X =$ some expression composed only of the terms A, B, and C." Lambert's solution is as follows: "Given $x/a = b$ and $a/x = c$, we can conclude that $ax = a - c = x - b$. Consequently (*folglich*), $x = a + b - c$." Lambert's solution is correct. Indeed, adding the equations $X/A = B$ and $A/X = C$ termwise, we obtain

$$X/A + A/X = B + C.$$

But, by (2) above, we can replace the sum $X/A + A/X$ with the equivalent sum $A + X$. Therefore, we have $A + X = B + C$. Adding A to both sides of this equation and taking into account $A + A = 0$, we obtain $X = A + B + C$. From the premise $A/X = C$, it follows that all C are A, that is, $C = CA$. Therefore, in the equation $X = A + B + C$ we can substitute CA for C, which yields $X = A + B + AC$, or $X = A(1 + C) + B$, or (since in Lambert's system, with its addition modulo two, $1 + C = 1 - C$) finally $X = (A - C) + B$. This coincides with Lambert's result.

By formulating a large number of problems analogous to the preceding, Lambert himself established the beginnings of a study of the solution of so-called logical (actually Boolean) equations. This theory was later to undergo intensive development in the works of G. Boole, E. Schröder, and P. Poretskiy.

Lambert also used an extremely interesting system of notation, which includes as symbols the constants 1 and 0; 1 is interpreted as the presence, and 0 as the absence of a given quality. He denoted constituent units in the form ABC, $AB0$, $A0C$, $0BC$, . . ., where the sign 0 in the corresponding place indicates that this place should be occupied by the appropriate letter (taken in alphabetical order) with a symbol of negation over it. Thus, Lambert determined the number of constituent units to be 2^n, where n is the number of letters denoting elementary classes.

Evidently following Leibniz' method, Lambert proposed a geometric interpretation of syllogistic functors, obtained by placing each term in correspondence with a straight-line segment. The proposition "Every *A* is *B*" was expressed by Lambert as a system of two unequal line segments, with the smaller segment placed

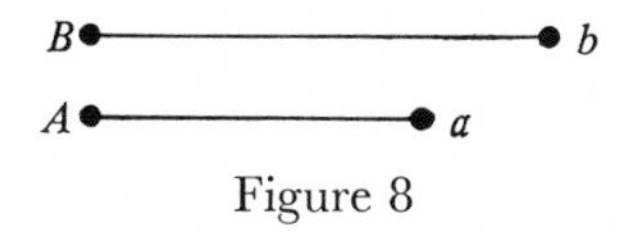

Figure 8

below (or above) the larger (see Figure 8). The line segment *Aa* is shorter than the segment *Bb* and lies under it.

The proposition "Not a single *A* is *B*" is represented by the diagram in Figure 9, where the line segments *Bb* and *Aa* are placed

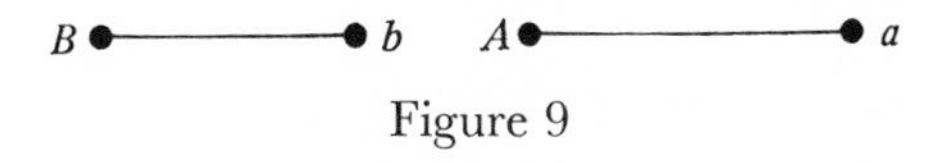

Figure 9

one outside the other along the same straight line. Between the right-hand end of the segment *Bb* and the left-hand end of the segment *Aa* there is an empty interval.

The proposition "Some *A* are not *B*" is represented by the Lambert diagram shown in Figure 10, where the thick portion of the

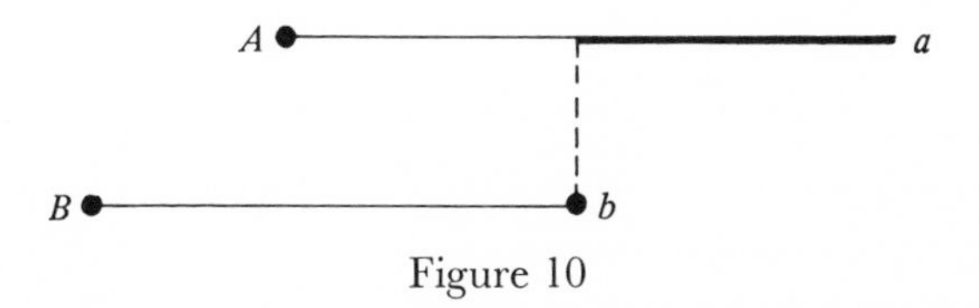

Figure 10

line segment *Aa* represents that part of the volume of the subject of the partially negative statement which lies outside of its predicate.

The proposition "Some *A* are *B*" is expressed by the diagram in Figure 11, where the thick portion of the line segment *Aa* symbolizes

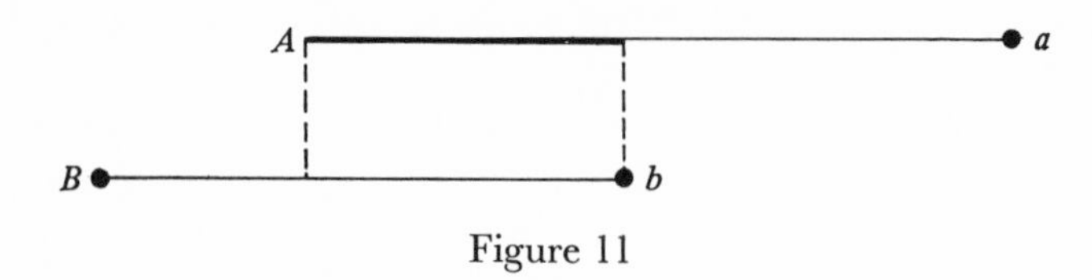

Figure 11

that part of the volume of the subject of the partially affirmative statement which is included within the limits of the predicate.

The syllogism "All B are A; some C are not A; therefore, some C are not B" can be expressed in a diagram as shown in Figure 12.

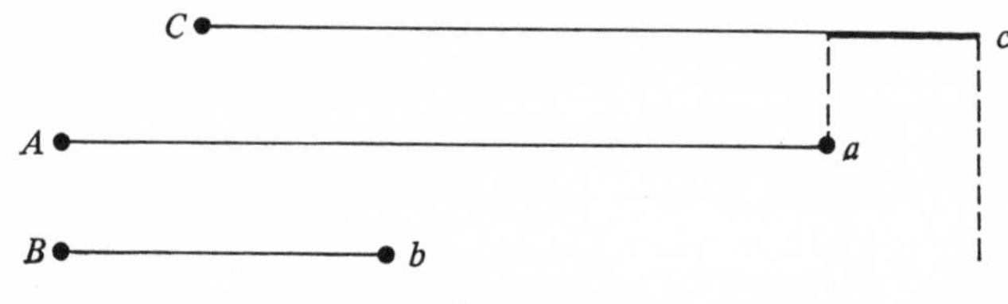

Figure 12

Applying the language of his calculus to formalization of traditional syllogistic theory, Lambert, first of all, quantified the predicate of a statement by using undetermined coefficients. The proposition "Every A is B" is given by the equation $A = mB$, where m is an undetermined coefficient. Later, the reader can verify that this is exactly the way George Boole quantified the predicate. The syllogism *Barbara* is given by the system of equations

$$\begin{array}{r} B = mA \\ C = vB \\ \hline \text{Consequently, } C = vmA. \end{array}$$

For the term B in the second premise, we substitute its value from the first premise (that is, the expression mA); since, however, the product of two undetermined coefficients is again an undetermined coefficient, we can set $vm = \Phi$ in the conclusion, where Φ is an undetermined coefficient. Hence, we finally obtain (Reference 238, vol. 1, pp. 103–104):

$$C = \Phi A.$$

A universally affirmative proposition (Every A is B) was also written by Lambert in two other forms: $A = B$ and $A > B$ (here the sign $>$ indicates that the predicate B is included in the subject A in the sense of being a feature of A). The conversion of these symbols yields universal and partial propositions, respectively (Reference 238, vol. 1, p. 93). Analogously, a partially affirmative proposition can be represented in two forms, depending on whether or not its conversion gives rise to a universal or a partial proposition; for these two cases, Lambert used the representations $mA > B$ and $A < nB$, depending on whether the subject was part of the predicate or the predicate part of the subject (Reference 238, vol. 1, p. 94).

The corresponding subdivisions were made by Lambert for universal and partial negative statements as well.

A large number of the modes of syllogisms deduced by Lambert naturally differ in form from the Aristotelian type.

It would not be an exaggeration to say that Lambert's theory of quantification actually contains all the basic results of the studies in quantification made by W. Hamilton in the nineteenth century.

In creating his generalized syllogistic theory, Lambert was well aware that it could not pretend to embrace all possible forms of scientific syllogisms. Thus, in the second volume of *Neves Organon*, Lambert pointed out the following conclusion from two partial premises as an example of one that cannot be expressed by a syllogistic formalism:

"$\frac{3}{4}$ of A are B; $\frac{2}{3}$ of A are C, therefore, some C are B"
(Reference 73, vol. 2, p. 355).

This argumentation is correct since the sum of $\frac{3}{4}$ and $\frac{2}{3}$ exceeds one by $\frac{5}{12}$. Therefore, there must exist at least $\frac{5}{12}$ of A that belongs simultaneously to B and C. In fact, we can replace the expressions $\frac{3}{4}$ and $\frac{2}{3}$ by the simpler term *most*, and we will still obtain the same conclusion. This latter fact was later carefully analyzed by the British nineteenth-century logician August De Morgan.

Finally, the last (but very significant) aspect of Lambert's logical system is his logical analysis of relations; this we find in his

Architectonics as well as in his *Logical and Philosophical Articles* (Reference 238, vol. 1, pp. 19–20 and Reference 240, vol. 1, p. 82). According to Lambert, the difference between a relation and a class, from the formal point of view, is that for the latter the law $X \cdot X = X$ holds, whereas for the former it does not necessarily hold (Reference 234, p. 96). Lambert sometimes called a relation a "metaphysical sign" (Reference 238, vol. p. 19).

Lambert often used formulas of the type $i = \alpha : : c$, where i and c denote subjects (*individuums*), α is the operator of a function, and the symbol : : is the sign of a relation. If we replace α with R, and the symbol : : with parentheses that enclose the variable(s), we see that Lambert's formula is equivalent to the expression $i = R(c)$ or, in words, "i is R of c."

This interpretation, in our opinion, allows us to correctly understand many of Lambert's equations. For example, we can understand Lambert's assertion that the equation $N : : b = M : : b$ does not necessarily imply the equation $N = M$. Indeed, the fact that certain specific values of two functions coincide [for example, $R(b)$ and $S(b)$] certainly does not imply that these functions (R and S) are identical. For example, in trigonometry tan 45° = cotan 45° (that is, R = tan, S = cotan, b = 45°), but this does not imply that tangent and cotangent are identical trigonometric functions.

Among the relations, Lambert singled out a special group having the property of very large volume (category): "among concepts of the form $M = P : : B$ we encounter concepts which are very wide in character. Such concepts are the following: reasons, actions, means, aims, bases, forms, types" (Reference 240, vol. 1, p. 82). However, we will not deal with Lambert's analysis of categories since it goes beyond the limits of proper logical problematics, relating rather to scientific methodology.

The difficulty of applying Lambert's formalism to the analysis of problems in traditional logic relates particularly to the fact that his system is "double-layered": it was conceived to be equally applicable to operations with classes and those with relations. Lambert was not able to make the further subdivision of the propositional calculus so as to produce individual self-sufficient fragments of his theory. In analyzing the history of the origins and development of mathematical logic, we can once again appreciate the truth of Karl Marx's observation that, in the initial phase of its evolution, a science sometimes occupies itself with "construction of the upper stories of a building before the foundation has been laid" (Reference 242, p. 46). In this sense, the early history of mathematical logic bears many traits similar to those noted by Marx in his analysis of the genesis of political-economic theory.

As was common in that era, Lambert carried on a lively correspondence with his scientific colleagues: Euler, Kant, Ploucquet, von Holland, and others. His letters to Euler concerned mathematical and physical questions; those to Kant, philosophical questions. Lambert's correspondence with George Johann von Holland, the German mathematician and student of Ploucquet, is collected in the first volume of J. H. Lambert's *Deutscher Gelehrter Briefwechsel, herausgegeben von J. Bernoulli* (Berlin, 1781). Von Holland, in his own time, was known as the author of the work *Abhandlung über die Mathematik, die allgemeine Zeichenkunst und die Verschiedenheit des Rechnungsarten* (Tübingen, 1764). As is obvious from even a single chapter of this work, von Holland defines mathematics as a universal symbolic theory, amenable to computational treatment. Here the influence of Leibniz' "universal characteristic" can be clearly felt. Von Holland compared Ploucquet's calculus with Lambert's method and also developed his own system of notation for formalizing syllogisms; he discussed this system in a letter to Professor Koestner, published in 1764 in the journal *Göttinger Anzeigen von gelehrten Sachen.*

In a letter of April 9, 1765 to Lambert, von Holland proposed the following interesting propositional scheme for obtaining the fundamental propositions of syllogistics:

$$\frac{S}{r} = \frac{P}{\pi}, \tag{I}$$

where S and P are symbols for the subject and predicate of a statement, respectively, and r and π are parameters taking the values 1, ∞ (infinity), and v (an undetermined positive whole number). By substituting these values in I, we obtain various forms of propositions. The following substitutions are possible:

1. $r = 1, \pi = \infty$;   4. $r = \infty, \pi = 1$;   7. $r = 1, \pi = 1$;
2. $r = 1, \pi = v$;   5. $r = v, \pi = 1$;   8. $r = \infty, \pi = \infty$;
3. $r = v, \pi = \infty$;   6. $r = \infty, \pi = v$;   9. $r = v, \pi = v$.

The symbol $\frac{x}{\infty}$ is considered equivalent to the expression "not-*x*." Substituting 1–9 successively in I, von Holland obtains from it three different forms for a universally affirmative proposition (XaY), three for a universally negative proposition (XeY), two for a partially affirmative (XiY), and one for a partially negative (XoY) proposition. (The letters *a*, *e*, *i*, and *o* serve to denote the corresponding universal and partial propositions of syllogistics). We obtain the following forms of statements, respectively (the bar – denotes logical negation):

1′. $S = \bar{P}$ (literally: only S are not-P),
2′. $vS = P$ (some S are P),
3′. $S = v\bar{P}$ (no S are P),
4′. $\bar{S} = P$ (only not-S are P),
5′. $S = vP$ (every S is P),
6′. $v\bar{S} = P$ (some not-S are P),
7′. $S = P$ (all S are P and, conversely, all P are S),
8′. $\bar{S} = \bar{P}$ (all not-S are not-P and, conversely, all not-P are not-S),
9′. $vS = vP$ (some S are some P).

It is easy to see that 7′ is equivalent to 8′, and 1′ is equivalent to 4′. Von Holland continued further to the formalization of the different forms of Aristotelian syllogisms. In particular, the mode *Barbara* is converted to the form

$$\text{Premises: } H = \frac{M}{r};\quad E = \frac{H}{\pi}. \qquad \text{Conclusion: } E = \frac{M}{r\pi}.$$

Here M plays the role of the major term; E, the minor term; H, the middle term.

Returning to consideration of von Holland's propositional

scheme, we should note that essentially it does not take into account the inverse logical operation of division. The terms S and P are not divided by the parameters r and π but are multiplied by $1/r$ and $1/\pi$, respectively. In other words, the operation of division is shifted onto the undetermined coefficients of S and P, so that scheme I can be written in the form

$$\frac{1}{r}S = \frac{1}{\pi}P. \tag{II}$$

Setting $1/r$ equal to m and $1/\pi$ equal to m' in II, we obtain

$$mS = m'P, \tag{III}$$

where m and m' are indeterminants. In III, it is natural to consider that the range of values of the indeterminants is contained in the interval from 0 to 1. In particular, this is the way that Venn (Reference 328, p. 157) treated von Holland's propositional scheme, correctly noting its similarity to George Boole's corresponding formula, which is equivalent to Representation III.

Von Holland's results, often occurring only in the context of a largely random logico-algebraic juxtaposition, have less meaning than the corresponding achievements of Lambert.

3. Intensional Logical Calculuses of the Last Decade of the Eighteenth Century (the Results of S. Maimon and G. F. Castillon)

The younger generation of philosophical disciples of C. Wolff "took great pains" to see that the logico-mathematical ideas of Leibniz were forgotten. At the same time, Lambert and Ploucquet's complicated logical constructions, with their unusual nature and somewhat unwieldy apparatus, frightened readers brought up in the spirit of traditional logic

Prominent philosophers and, occasionally, well-known mathematicians blurred the fact of the existence of any important applications of the new logical teachings. Hegel's polemic against Ploucquet had a serious negative influence. Whereas Kant limited himself to the assertion that the Aristotelian logical formalism could not be surpassed, Hegel attempted to mock the very idea of the mathematization of logic.

The position of the two most important philosophers of that time (especially Kant's) could not help but undermine the confidence of some scientists in the emerging logico-mathematical theory. Those thinkers who, nevertheless, had the boldness to again return to the development of logico-algebraic ideas now relied on their own names, preferring not to cite either Lambert or Ploucquet. Thus, the Jewish philosopher Salomon Maimon (1753–1800), in proposing his own logical calculus in 1794, wrote: "I do not know whether anything has been done before me in the realm of the symbolic logical art. Lambert's method of lines is completely different from my method; also, what Ploucquet supposedly did in this field is unknown to me" (Reference 243, par. XXVII). A similar position was taken by another pioneer in mathematical logic, G. F. Castillon (1747–1814).

Of course, the originality of the results of Maimon and Castillon cannot be doubted; however, their results (in fact, somewhat weaker than those of Lambert and Ploucquet) hardly gave them any justification for disowning the achievements of their illustrious predecessors.

Salomon Maimon was born in 1753 into a poor rabbi's family living in a remote Lithuanian village near the small town of Neswij on one of the estates of Prince Radziwill. From birth Salomon was known by the name Ben Joshua; the name Maimon was given to him by his Berlin friends, probably out of respect for the philosopher Maimonides.

Maimon's father was first a Talmudist, then a tradesman, later a tenant farmer, and finally a part-time teacher. When Salomon was six, his father decided that the time had come for him to begin the study of the Bible. The unusual quickness of the youngster was somewhat perplexing to his teacher, but a solution was quickly found: the seven-year-old Maimon was sent to a school for Talmudists. To ordinary children, this special school was very difficult. However, Maimon's unusual ability and memory made him at age eleven a serious competitor for any of the prospective preachers. Maimon's father hurried to carry out the usual "conclusion" for this situation as well: as a result, Maimon at fourteen was not only a married man but the father of a son. From the beginning, marriage brought him only grief.

Maimon's Lithuanian environs did not suffer from a surplus of books, and he had to subsist by studying the Talmud. Fortunately for him, he was soon able to obtain Moses Maimonides book *A Guide to the Perplexed*; this tome, an encyclopedia for rabbis, had been carefully studied by Leibniz himself in his time. Later, several German books on medicine also came into Maimon's hands.

Only with great difficulty was Maimon able to provide for his family: the meager wages of a tutor were far from adequate. It began to appear that Maimon would be drowned in a quagmire of commonness and that he would be unable to develop fully his rich natural gifts. But then the twenty-five-year-old Maimon set out for Germany, the land of books and enlightenment, to "study medicine and perhaps other sciences" (Reference 244, p. 404).

Finding himself among the students of Koenigsberg, Maimon at first felt very constrained; he often fell into comic situations. The difficulty arose not only because of his shyness but also because of language problems. At first, he expressed himself in his own peculiar mixture of Jewish, Polish, German, and Lithuanian, or, more precisely (as Maimon himself shrewdly observed), his language "was put together from Hebrew, German, and Russian plus the grammatical errors characteristic of each of these languages." (Reference 244, pp. 404–405.)

Maimon could not stay very long in either Koenigsberg or Berlin. He was forced to leave Berlin because of the condemnation of an Orthodox rabbi to whom he had presented his commentary to Maimonides' *A Guide to the Perplexed.*

Maimon's life became a series of wanderings. Three years were devoted to the study of pharmacy. He became very poor and was never able to attain any kind of steady material situation. A philosopher, however, never particularly strives for this. Maimon's conduct throughout his life reminds one of Spinoza and is in sharp contrast to that of some of his philosophical colleagues, who were not burdened by the problem of moral perfection.

Only after many wanderings and painful experiences did Maimon return to Berlin. There he made the acquaintance of the Leibnizian Moses Mendelssohn (1729–1786), who, in his critique of religious Scholastic philosophy, adhered to some of the ideas of the materialist Spinoza's *Theological and Political Treatise*. Under Mendelssohn's guidance, Maimon set to work on the philosophical works of Wolff, Spinoza, Locke, and Kant. Later, he defended Spinoza against the polemic attacks of Mendelssohn and the Wolffians.

In 1794 Maimon published his fundamental work *Experiment in New Logic*; it was this work that introduced the term "mathematical logic" (*matematische Logik*) (Reference 243, 1st ed., p. 245) although in the sense of a general study of quantities (*Grössenlehre*). Maimon felt his work was in keeping with the idea of a universal mathematics expressed by Descartes and Leibniz. His contemporaries considered Maimon a representative of "logical formalism."

Maimon died November 22, 1800. His autobiography (1792) is of special interest because of its depiction of the everyday life of Russian-Polish Jews in the eighteenth century.

Maimon gained renown in philosophical circles because of his sharp-witted critique of Kant. In general, it must be noted that all of the pioneers of mathematical logic who lived in Kant's time were extremely unsympathetic to his philosophy, which asserted that formal logic was hopelessly stagnated.

By attributing apodictic truth exclusively to mathematical statements, Maimon, unlike Kant, intended to derive meaning not from synthetic statements (in the Kantian sense) but from analytic forms of thought. In Maimon's philosophical works, commentary on and criticism of Kant is closely linked to the presentation of Maimon's own philosophical theory. In his critique of Kant, he was joined by Kant's former pupil Jakob Sigmund Beck (1761–1840). Maimon refuted Kant's teachings about categories, demanded the inclusion of the concept of spiritual pleasure in any ethical system, and proclaimed as the highest principle of practical ethics *the gratifying feeling of one's own dignity*. As a whole, Maimon's philosophy flows into the general stream of early Post-Kantian metaphysics.

Of Maimon's philosophical and logical works, we should note the following: *Versuch über die transzendentale Philosophie: mit einem Anhang über die symbolischen Erkenntniss und Anmerkungen von S. Maimon* (Berlin, 1790); *Über die Progressen der Philosophie veranlasst durch die Preisfrage der Königlichen Akademie zu Berlin für das Jahr 1792*; *Was hat die Metaphysik seit Leibniz und Wolff für Progressen gemacht?* (Berlin, 1793); *Notes to The Works of Bacon of Verulamium*, translation by G. W. Bertholdi (Berlin, 1793; only one volume was published); *Streifereien im Gebiete der Philosophie* (Berlin, 1793); *Versuch einer neuen Logik oder Theorie des Denkens nebst angehängten Briefen des Philaletes an Aenesidemus* (Berlin, 1794); *Die Kategorian des Aristoteles. Mit Anmerkungen, erläutert und als Propädentik zu einer Theorie des Denkens dargestellt* (Berlin, 1794); *Kritische Untersuchungen über die menschliche Geist oder das höhere Erkenntnis- und Willensvermögen* (Leipzig, 1797).

For literature on Maimon's logical and philosophical views see References 244–254.

In the Introduction to Reference 243, Maimon emphasizes that the logical method presented by him represents only a formal instrument "not dealing with the operation of thought in itself" (Reference 243, p. XXVIII). This remark is directed against any possible psychological interpretation of his calculus.

As a whole, Maimon's calculus consists of a formalization of syllogistics plus several theorems relating to propositional logic. Below, for the purpose of clarity, we will compare the notation of Maimon's language with the notation for syllogistic expressions in the language of J. Lukasiewicz.

As a symbolization of the functor "is," Maimon used the sign $+$. For him, the expression $ax + b$ denotes Aab (every a is b), where the letter x corresponds to the universal quantifier. In Maimon's calculus, the expression $an + b$ denotes Iab (some a are b), where the letter n corresponds to the existential quantifier. The statement Oab (some a are not b) is written in the form: $-ax + b$ (literally: not all a are b). The statement Eab (not one a is b) is represented by Maimon in the form $ax - b$. Strictly speaking, it should be written as $ax + (-b)$, but in such cases Maimon adopted the convention of omitting the $+$ sign. To avoid confusion, however, we should consider the expression $ax - b$ only an abbreviation for the expression $ax + (-b)$.

The expression $ax + a$ denotes Aaa; the expression $an + a$ denotes Iaa. Let us note some of the theorems formulated in Maimon's calculus:

1. The expression $ax - (-a)$ is equivalent to $ax + a$. This theorem corresponds to the equivalence $Aa\bar{\bar{a}} \rightleftarrows Aaa$. To prove this, we apply the law of double negation.

2. If $ax = a$ is true, then $an = a$ is true. This theorem obviously corresponds to the implication

$$(\forall x(A(x) = A(x))) \rightarrow (\exists x(A(x) = A(x))).$$

3. From $an = b$, it does not follow that $ax = b$ since $x > n$. In other words, from $\exists x(A(x) = B(x))$, it does not follow that $\forall x(A(x) = B(x))$, since the existential quantifier is subordinate to the universal quantifier, but the opposite is not true.

4. The expression $ax - (-a)$ implies $(-a)x - a$. In other words, the implication $Aaa \rightarrow A\bar{a}\bar{a}$ holds.

5. $an + b$ implies $bn + a$; that is, we have $Iab \rightarrow Iba$.

6. If $a = b$, then $-a = -b$. This theorem relates to propositional calculus: $(A \equiv B) \rightarrow (\bar{A} \equiv \bar{B})$.

The unquantified affirmative proposition "*a* is *b*" was written by Maimon as $a + b$ or $ab + b$. This latter representation deserves special attention: it is of analytic character in the Leibnizian sense (that is, the content of the predicate b is already included in the content of the subject a). Thus, the thesis *a* is *b*, according to Leibniz and Maimon, is equivalent to the thesis (*a vel b*) is *b*. In other words, they assert that the proposition that a property b is inherent to some a is equivalent to the assertion that the property b is already contained in a sum of properties, a and b. Symbolically, we have

$$a \rightarrow b \qquad \text{is equivalent to} \qquad a \vee b = b,$$

where $\rightarrow$ should be understood in the sense of *is*. Maimon's expression $ab + b$, thus, means precisely the equation $a \vee b = b$. The omission of a symbol between letters (with the exception of quantifying forms of the type ax and an) indicated for Maimon the operation of addition of properties, which thus seems to be related to Ploucquet's "association of ideas."

In certain places (Reference 243), Maimon used strict disjunction, which he denotes by a vertical line: |. He wrote: "Mutually exclusive articles will be separated one from the other by a vertical line |, for example: $a + b \mid c \mid d$" (Reference 243, p. 60).

Maimon applied his calculus first to syllogistics, then to the theory of complex (compound) syllogisms. Later, M. Drobisch found exactly the same application for his own logical formalism.

It must be pointed out that when Maimon formalized classical Aristotelian syllogisms, he represented statements with forms in which symbols analogous to quantifiers are absent. It is easy to see that if he used his forms for quantifiers (of the type $Sx + P$), he would not meet with much success. In fact, it is difficult to see how one can obtain from the premises

Π_1. $Sx + P$
Π_2. $Rx + S$

the conclusion $Rx + P$ (that is, how to formalize mode *Barbara*). Maimon escaped from this difficulty by applying Leibniz' theory about the analytic nature of statements. He wrote the statement $Xx + Y$ (all X are Y) as $XY + Y$, thus replacing the lower-case x with Y; we should understand this as $X \vee Y = Y$. Let us represent Π_1 with $S \vee P = P$, and Π_2 with $R \vee S = S$. Now we replace the term S in Π_1 with its value taken from Π_2: $(R \vee S) \vee P = P$, equivalent to $R \vee (S \vee P) = P$. In this latter equation, we substitute P for the expression $S \vee P$, in accordance with proposition Π_1. We obtain $R \vee P = P$ or, in Maimon's notation, $Rx + P$. The other Aristotelian modes are formalized in a similar way.

On the question of axiomatization of syllogistic theory, Maimon treated the axiom of the first figure as stating the transitivity of the copula +, and the axioms of the second and third figures as parallel to the arithmetic axiom which says that two quantities, each separately equal to a third, are equal to each other.

In conclusion, Maimon's logical calculus can be characterized as a consistent attempt to extend the Leibnizian intensional interpretation of statements to the case of the formalization of the classical theory of syllogisms. The peculiarity of Maimon's formalism is the attempt to explicitly symbolize the universal and existential quantifiers.

The intensional logical calculus of the German philosopher G. F. Castillon (1747–1814) is also worthy of our attention. The son of a well-known mathematician (J. Castillon, 1708–1794) G. F. Castillon held the position of professor of philosophy at the Berlin military academy. He also taught psychology and logic. In 1786 he was chosen a member of the Berlin Academy of Sciences, serving as the head of its philosophical section from 1801 to 1812. The logical ideas of Castillon were formulated under the strong influence of Lambert's work.

Since, like Lambert and Segner before him, Castillon had a wide understanding of the essence of algebraic calculuses (as not reducing only to the generalization of numerical arithmetic), he was quite familiar with the attempts to apply them to the formalization of logic. He attempted to introduce into logic a minimal number of algebraically treatable functors, which would play the role of logical associations.

An intensional formalization of syllogistics was presented by Castillon in the article "Memoire sur un novel algorithme logique" (November 3, 1803). This article presents the basic concepts and operations of Castillon's calculus. "Every concept or notion," he said, "is denoted by a different letter. Synthesis or composition of concepts is denoted by algebraic addition, that is, by use of the + sign; analysis or decomposition of concepts is denoted by algebraic subtraction, that is, by the sign −. Inference (*jugement*) is denoted by using an equation; to the left of the equal sign we place the subject; to the right, the predicate. In the case of a negative statement, the predicate is preceded by a negation sign. This predicate is joined by a + sign to the other concept since (for the most part) the concept of the subject is not exhausted simply by the assertion of the presence in it of some feature (in the case of an affirmative proposition) or of the necessary absence in it of that feature (in the case of a negative proposition)" (Reference 255).

The statement "every S is P" was written by Castillon in the form $S = P + M$, where P is interpreted as a "defined" term, M an undetermined one. This is his representation of the statement *SaP*.

The statement *SeP* (not one S is P) is written as $S = -P + M$. The statement *SiP* (some S are P) is represented by the equation $P = S - M$, and *SoP* (there exist S which are not P) is given by the equation $-P = S - M$.

Castillon formalizes the syllogism *Barbara* in the following way.

$$\begin{array}{l} P = T + N \\ T = S + Z \\ \hline P = S + (N + Z) \end{array}$$

Here N and Z are undetermined terms. In the conclusion, we can set $N + Z = W$, where W is also an undetermined term; then we finally obtain $P = S + W$.

In analyzing Castillon's calculus, C. I. Lewis notes that the difficulty with the introduction of subtraction in intensional logic is eliminated in Castillon's calculus "only by means of the constant process of combination of undetermined terms; this conceals the fact that the operations + and − used by Castillon are in actuality not universal" (Reference 257; cited in Reference 326, vol. 1, p. 87).

In our view, the basic difficulties in this calculus stem from the fact that the "undetermined" term M is assumed, actually, by the related quantifier: either the existential quantifier (as in the formula $S = P + M$) or the universal quantifier (as in the formula $-P = S - M$). With this understanding, we can construe Castillon's operation + as the symbol for conjunction, − as the symbol for inclusive disjunction, and the compound symbol $= -X$ (where X is a term) as an abbreviation for the expression "equals not-X."* Castillon's proposition $S = P + M$ can be expressed by the formula $\exists v \forall x (S(x) = P(x) \,\&\, v)$, where v plays the

* Analogously, Castillon's notation $-X =$ corresponds to the expression not-x equals. Whereas in formulas of the form $X - Y$ the minus sign plays the role of inclusive *or*, in the notations $-X =$ and $= -X$ it plays the role of the logical negation functor. Strictly speaking, Castillon should have chosen a distinct symbol for this.

role of M. The sense of the statement $-P = S - M$ is rendered by the formula

$$\forall v \exists x (P(x) = -(S(x) \vee v)).$$

The proposition $S = -P + M$ reduces to

$$\exists v \forall x (S(x) = \bar{p}(x) \,\&\, v),$$

and $P = S - M$ reduces to

$$\forall v \exists x (P(x) = S(x) \vee v).*$$

Thus, the special characteristic of Castillon's syllogistic formalism (in its precise description in contemporary terms) consists in the following fact: in it, universal propositions are expressed by use of existential operators on (undetermined) predicates, while universal quantifiers are taken on individuums; on the other hand, partial statements are expressed by use of universal operators on (undetermined) predicates, while existential quantifiers are taken on individuums.

However, since Castillon himself was unable to formally express this duality of the term v in his calculus, he encountered serious difficulties in his attempts to apply his algorithm to analysis of systems of substantial statements, in particular to synthetic random propositions (Reference 256).

* In this book, the symbol $\exists$ represents the existential quantifier, and the symbol $\forall$ represents the universal quantifier.

CHAPTER FOUR

Forerunners of the Algebra of Logic of George Boole

1. The Development of Mathematical Logic in the First Half of the Nineteenth Century

In his work *Versuch über die combinatorische Methode: ein Beitrag zur angewandten Logik und allgemeinen Methodik* (Dresden, 1811), the German logician Christian August Semler emerged as a forerunner of the English logicians George Boole and Stanley Jevons. He persistently demanded enumeration of all possible combinations which could be formed from the specified terms for classes. First Semler exhibited all such combinations of classes; then he eliminated the various classes or combinations of classes that contradicted the given premises.

The text we present will demonstrate that Semler's combinatorial method is identical to W. S. Jevons' so-called empirical method. Semler writes: "For this example, let us form the 56 couplings which can be constructed from the concepts *a*, *b*, *c*, *d*, *e*, *f*, *g*, *h*, taken as pairs without regard to order. Now we note that the concepts *a* and *b*, as well as the concepts *b* and *d*, cannot both be realized without contradicting the given premises. In this way, we can take all the possible couplings and eliminate the impossible ones; then we gather the remaining ones in a separate sample (in our example, ten such couplings remain)" (Reference 258; cited in Reference 254, p. 415).

Thus, we take *p* terms for classes and form from them, through the operations of intersection and union, the *Qp* possible couplings; then we reduce this number *Qp* by using the given *m* premises. The latter are written in the form of assertions on the incompatibility of certain terms, which implies that certain combinations of classes are impossible.

Semler studied the origins of the logic of relations in Ploucquet's work very carefully; in particular, he noted that in the case of certain relations the law of commutativity cannot be considered universal (Reference 258). In the latter case, Semler perceived one of the formal differences between the calculus of relations and the logic of classes. He was conversant not only with Ploucquet's logical results but with those of Lambert as well.

The French astronomer and mathematician J. D. Gergonne (1771–1859) also made a significant contribution to the development of the logical calculus of classes. Gergonne was editor-in-chief of the journal *Annales Mathematiques pures et appliques*. By 1820–1821, he had already made a very thorough study of the dual concept of nets of conic sections. In his works of 1827–1829, he treated the principle of duality in projective geometry from a more general point of view. He also worked on combinatorial analysis and did some studies on inversions. As a supplement to Laplace's theory, he studied the solution of systems of linear equations with a large number of unknowns. Gergonne widely applied the language of analytic geometry to the treatment of questions concerning algebraic curves and second-order surfaces. In 1810–1811 he extended the theory of polar curves to second-order surfaces.

The following of Gergonne's works present results of logical studies: "Essai de dialectique rationelle," *Annales des Mathématiques pures et appliqueles*, no. 7 (1816–1817); "Essai sur la théorie de definitions." For Gergonne's logic, see also References 259 and 368 (pp. 212–213).

The French mathematician considered five basic relations between the classes a and b, and used these relations to express the classical Aristotelian proposition denoted by the letters A, E, I, and O. The relations between the classes a and b used by Gergonne were the following:

1. complete coincidence (equivalence);
2. left-sided inclusion (a is contained in b);

3. partial intersection;
4. right-sided inclusion (b is contained in a);
5. nonintersection (a and b are disjoint).

The relations are graphically illustrated in Figure 13.

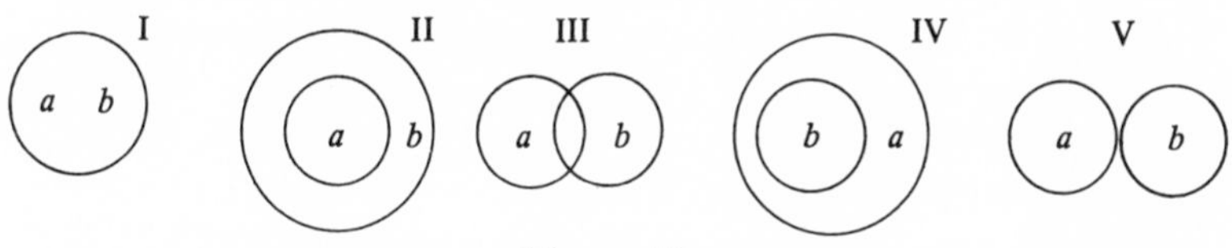

Figure 13

Gergonne, like Aristotle, presupposes that the classes a and b are not empty; that is, he assumes that: $(a \neq 0)$ & $(b \neq 0)$, where 0 is the symbol for the empty set, & denotes conjunction, and the symbol $\neq$ denotes the relation of inequality.

Statement A (that is, $a \leqslant b$) is denoted by Gergonne as $(a = b) \vee ((a \leqslant b) \,\&\, \overline{(b = a)})$, where $\leqslant$ corresponds to the functor *is*, $\vee$ is the disjunction sign, and $-$ is the functor of logical negation.

Statement I is represented by equivalence, denoted by the symbol $\equiv$:

$$(\overline{a \cdot b = 0}) \equiv ((a \leqslant b) \vee (b \leqslant a) \vee (\overline{ab = 0}) \,\&\, (\overline{a \leqslant b}) \,\&\, (\overline{b \leqslant a}))$$

where the symbol $\cdot$ denotes set intersection.

The statement O is expressed as

$$(\overline{a \cdot \bar{b}} = 0) \equiv ((a \leqslant \bar{b}) \vee (\bar{b} \leqslant a) \vee (\overline{a\bar{b} = 0}) \,\&\, (\overline{a \leqslant \bar{b}}) \,\&\, (\overline{\bar{b} \leqslant a})).$$

Finally, the statement E is expressed as $ab = 0$ (the intersection of the classes a and b is empty).

Let us now express A, E, I, and O in abbreviated form, by taking into account the graphical representations I–V (Figure 13) between a and b. Then we obtain the following group of equivalences:

1. $A \equiv \mathrm{I} \vee \mathrm{II}$;
2. $E \equiv \mathrm{V}$;
3. $I \equiv \mathrm{I} \vee \mathrm{II} \vee \mathrm{III} \vee \mathrm{IV}$;
4. $O \equiv \mathrm{III} \vee \mathrm{IV} \vee \mathrm{V}$.

In the actual notation for his relations, Gergonne was guided by mnemonic considerations. Thus, Gergonne denoted the relation of nonintersection with the letter H (the first letter of the French word *hors*, meaning "outside of"). The relation of partial intersection was denoted by X, identity by the letter I. These three letters can all be reversed without changing their form. In mnemonic language, this means that the relations represented by these symbols are symmetric.

The German logician A. D. Twesten (early nineteenth century) was a scientist who went far beyond the bounds of traditional logical descriptions. Many of his ideas are in harmony with the logical-mathematical conception in the study of the laws of rational thought. Certain of his other ideas recall Gergonne's analysis of the basic relations between classes. Twesten's essential logical results are most characteristically represented in the following two works: *Die Logik, insbesondere die Analytik* (Schleswieg, 1825) and *Grundriss der analytischen Logik* (Kiel, 1834).

According to Twesten, logic is the science of application of the laws of identity and contradiction (Reference 260, p. 4). In this sense, logic is analytic. In a wider sense, the subject of logic can also include the study of the synthetic aspects, linked to a discussion of such problems as the construction of concepts and statements, and the structure of knowledge as a whole (*Erzeugung oder Zusammensetzung unserer Begriffe, Urtheile und Erkenntnisse*; Reference 260, pp. 4–5).

Some historians of logic, for example, P. Leikfeld (Reference 262), consider Twesten a Kantian because he felt logic concerns only formal (and not material) truth criteria. However, by this token, one would have to consider as Kantians all the representatives of formal logic (beginning with Aristotle), and this would be absurd.

In the theory of propositions, Twesten sided with the conception of the subject as included in the predicate (in quantity).

In his studies on syllogisms, the German logician occupied himself, in particular, with the development of the theory of syllogistic chains (*Schlussketten*; Reference 260, p. 133 ff.). He intended to consider these chain conclusions as continued figures of classical syllogisms. Twesten ordered syllogistic chains with respect to four possible places, which could occupy the greater and lesser terms in the first upper and last lower premises, on the model of the four figures of the syllogisms of school logic. First he studied those chains in which each subsequent syllogism has only one syllogism directly preceding it, the conclusion of that syllogism being one of its premises. Employing the terminology of A. L. Subbotin (Reference 262), we could say that here Twesten dealt with so-called "linear chains" of syllogisms. Later, he moved on to the characteristics of somewhat more complicated syllogistic chains, namely those in which subsequent syllogisms may have two syllogisms directly preceding them ("cascading chains" of syllogisms, in Subbotin's terminology).

Twesten also applied combinatorial analysis to compound syllogisms. His ideas in this area were later taken up by the Leipzig philosopher M. W. Drobisch.

As for the calculus of classes, Twesten, independently of Gergonne, also considered four basic relations between classes: identity, inclusion, partial intersection, and nonintersection (Reference 260, p. 30).

We will briefly note the characteristics of the formal logical symbolism of Twesten. The + sign is used to indicate the operation of addition of properties; that is, it is understood in the intensional sense. But since the intensional aspect is in a certain sense dual to the extensional interpretation (in the sense of volume or quantity) Twesten's + sign indicates the operation of multiplication (intersection) of classes. Extensional multiplication corresponds to intensional addition.

The remaining portion of Twesten's system of symbolic notation bore a clear influence of the logical symbolism of Lambert. Like Lambert, Twesten also devoted his attention to an analysis of the inverse logical operations of division and subtraction.

Another contributer to symbolic logic was the German mathematician K. F. Gauber (1775–1851). He is the author of *Scholae logico-mathematicae* (Stuttgart, 1829). Although this work is predominantly of a strictly mathematical character, it nevertheless contains many profound logical observations and generalizations.

In particular, Gauber established the logical principle sometimes called the law of closed (Gauber) systems (Reference 264, par. XII). The essence of this law is as follows. Suppose we are given three implications, $a \to b$, $c \to d$, $e \to f$, whose consequents (second terms) are mutually exclusive; that is, we have the following three conditions: $b \to \bar{d}$; $d \to \bar{f}$; $f \to \bar{b}$. At the same time, the disjunction of the antecedents (first terms) is equal to the identity constant:

$$a \vee c \vee e = 1,$$

where the symbol 1 denotes an expression which is always true. Then, the following implications are true:

$$\bar{a} \to \bar{b}; \qquad \bar{c} \to \bar{d}; \qquad \bar{e} \to \bar{f}.$$

In other words, in Gauber logical systems the truth of implications of the form $x \to y$ means that the converse forms $\bar{x} \to \bar{y}$ are also true.

We will prove Gauber's theorem for the case of two statements, after which it can be extended by induction to the case of three (or more) statements. Thus, we are given the following four conditions:

$$a \to b; \tag{1}$$

$$c \to d; \tag{2}$$

$$a \vee c = 1; \tag{3}$$

$$b \to \bar{d}. \tag{4}$$

We must prove that $\bar{a} \to \bar{b}$ and $\bar{c} \to \bar{d}$.

Proof: By the principle of syllogistics, we can, from the premises $a \to b$ and $b \to \bar{d}$, conclude that

$$a \to \bar{d}. \tag{5}$$

But, on the strength of Premise 3 we have

$$\bar{c} \to a. \tag{6}$$

From 5 and 6 and the principle of syllogism we obtain:

$$\bar{c} \to \bar{d}.$$

Thus, the second part of the theorem is proved. It remains to show that $\bar{a} \rightarrow \bar{b}$. From Premises 2 and 4 we obtain the implication

$$c \rightarrow \bar{b}; \tag{7}$$

from 7 and 3 we obtain $\bar{a} \rightarrow \bar{b}$. Gauber's theorem is proved.

The Czech scientist Bernard Bolzano (1781–1848) was born on October 5, 1781 in Prague. He was well-known as a mathematician, Utopian socialist, philosopher, and logician. W. Dubislaw (Reference 265) sees him as one of the pioneers of mathematical logic.

In 1796 Bolzano finished secondary school and joined the department of philosophy at Prague University. He was much influenced by the mathematical handbook of A. Koestner, who was Gauss' teacher. Bolzano completed his philosophy studies with honors in 1800 and then went on to finish his studies in the theology department.

By 1806, Bolzano was professor of religious philosophy at Prague University. However, the Austrian police state suspected him of political unreliability. His ideas about Utopian socialism, his passionate pleading of the abstract humanism inherited from Leibniz and Wolff, and his concern for the general welfare were all very displeasing to the authorities. As a result of the intrigues and denunciations of militant clerics, Bolzano was driven from the university in 1820 and was placed under police surveillance. The Jesuits also insisted that he be expelled from the holy order.

In 1843 Bolzano became seriously ill with pneumonia, from which he never fully recovered. He died on December 21, 1848.

Bolzano's fundamental work in the philosophy of mathematics, *Paradoxes of the Infinite* (1851), was published in Leipzig only after his death. The work was begun in the summer of 1847, when Bolzano was staying near the city of Melnik, and was completed in 1848.

In *Paradoxes* Bolzano became the first mathematician to note that an infinite set could be considered equivalent to certain of its subsets. Thus, for example, the set of positive whole numbers is equivalent to the set of positive even numbers, although it is clear that the set of even numbers is only a subset of the set of whole numbers. To establish such an equivalence, we need use only the function $f = 2n$, where n is a natural number. If we use the symbol $\updownarrow$ to denote the correspondence, we can represent this equivalence between the sets of whole and even numbers by using the diagram in Figure 14. The property of infinite sets noted by Bolzano was later used by C. S. Peirce to define precisely the concepts of finite

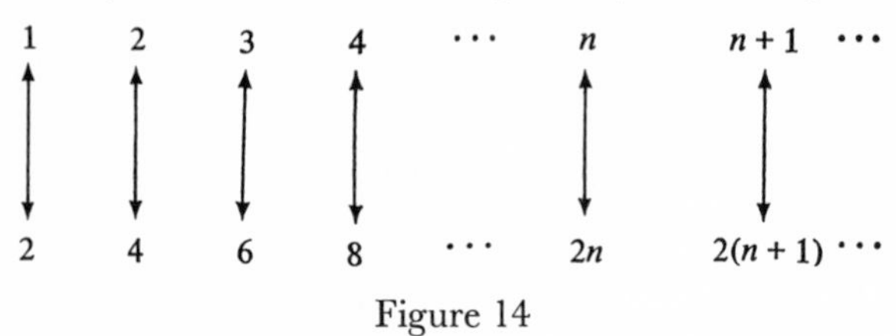

Figure 14

and infinite sets. Thus, Peirce defined a finite set as a set that is not equivalent to any of its proper subsets.

Bolzano actually laid the foundations for set theory. He also was a precursor of Cauchy, Cantor, and Weierstrass in the arithmetization of mathematical analysis.

Bolzano also had a very clear conception of the convergence of series. It was just recently discovered in his manuscripts that he was the first to give an example of a nowhere differentiable continuous function. And he did all of this before 1830! Many mathematical historians (for example, G. Vileitner, Reference 286) unanimously agree that a proper evaluation of Bolzano's ideas during his own time might have greatly accelerated the development of mathematics.

In philosophy Bolzano remained close to medieval realism and Platonism. He was especially interested in the problem of objective truth. According to Bolzano, to negate the existence of truth is impossible since such a negation would imply a clear contradiction. After all, even the negation of truth thereby projects a certain definite truth. And once there exists even one truth, there exist an infinite number of truths, for we can take any true proposition and form a new truth, that is, the statement that this true proposition is true, and so on. Bolzano treats truth as a property of a statement.

The reader familiar with the history of philosophy will easily note the similarity between Bolzano's proof of the existence of truth and the well-known argument of Plato in *Parmenides* dealing with the introduction of the concept of the natural numbers. If we have a unit, said Plato, then we have its existence, which differs from the unit itself just as a concept is different from its object. But *unit* and *existence* already make *two* objects. Thus, we have introduced the number two. Analogously, we can introduce three, four, and the rest of the natural numbers. A similar method of argument was also used by Bolzano (Reference 285, sec. 13, pp. 17–18), when he wanted to introduce the concept of an infinite set of logical degrees or levels (*Stufen*).

Bolzano differentiated the act of statement from its content (thought). The concept of the thought of a statement is linked by him to the term "thesis" (*Satz*). A thesis does not depend on whether or not someone proposes it. Using the ordinary terminology of German classical philosophy, we might say that according to Bolzano the thought of a statement is "thesis in itself," which, as existing, is described by the category *Sein* (being) but not *Dasein* (effective being). The concept of "truth in itself," according to Bolzano, should be somewhat analogous to the concept of "objective truth" (whether this is true would seem to us an entirely different question).

The logical results obtained by Bolzano are presented in his work *Wissenschaftslehre. Versuch einer ausführlichen und grössestens neuen Darstellung der Logik, mit steter Rücksicht auf deren bisherige Bearbeiter, herausgegeben von mehreren seiner Freunde.* (*Mit einer Vorrede von I. Ch. A. Heiroth*, 4 vols. (Sulzbach, 1837).

Bolzano developed his theories at a time when only an insignificant portion of the posthumously published works of Leibniz were available. Therefore, the possibility of Leibniz' direct influence on Bolzano is very small indeed. Certain points in Leibniz' metaphysics seemed doubtful to Bolzano. In particular, he did not agree with Leibnizian theodicy or with his thesis about the immutability of the monad.

A prominent place in Bolzano's *Wissenschaftslehre* (Reference 266) is occupied by a critique of the shortcomings of many traditional formal logical conceptions. For example, he criticized the formal-logic law on the inverse relation between the quantity (volume) and the content of a subject, pointing out many examples for which, in his opinion, this law does not hold. Bolzano considered the following example. Suppose we are given two concepts: "curved sphere" and "sphere." Their quantity is the same, but the content of the first is greater than the content of the second.

Let us consider another such pair of concepts: (1) people knowing all European languages and (2) people knowing all living European languages. Clearly, the quantity of (2) is greater than the quantity of (1); however, the law on inverse relations would require just the opposite. Generalizing his observations on these and analogous examples, Bolzano noted: "not every representation [this term is used here to mean concept — N.S.] which is subordinate to another is obtained by adding something to it" (cited in Reference 267, p. 103). Bolzano brought against traditional logic other, more serious, reproaches, related to methodological shortcomings in the structure of this discipline.

According to Bolzano, logic as a science must first of all accustom itself to distinguishing things with mathematical sharpness and to developing its concepts with mathematical exactness. Bolzano set down principles for an axiomatic construction of logic and made the initial attempts on the road to its realization.

Of special significance was the introduction by Bolzano of the so-called method of variation of representation, which he applied in order to establish the compatibility of propositions and their derivability one from another. For example, how can one establish a relation of subordination between propositions dealing with objects which are not real? Suppose, for example, that we have two such propositions: (p_1) a mountain of gold is large and (p_2) a mountain of metal is large. Let us replace the term "mountain" with the term "coin" in both (p_1) and (p_2); then proposition p_1 is subordinate to proposition p_2. The nature of this subordination is preserved under the inverse transformation of the term "coin" into the term "mountain."

We can see that in his method of variation of representation Bolzano was close to formulating the concept of a propositional function. Another aspect of this method consists in studying the logical structure of a statement on this basis.

In the theory of statements, Bolzano, along with Leibniz and

Kant, used the concept of "a statement in the abstract sense," which was different from the concept of "a statement in the traditional sense." The essence of this difference is explained by A. Church in Reference 268 as follows: if we have translations of one and the same sentence into many different languages, then we are dealing with various "statements in the traditional sense" but with only one "statement in the abstract sense."

In the theory of syllogisms, Bolzano gave the following definition of the concept *implication*: a set of formulas N follows from a set of formulas M (symbolically: $M\,Folg\,N$) if each model M is at the same time a model N.

Bolzano made wide use of topological illustrations of logical relations. He used rectangles to represent the volume of a term (concept). Bolzano operated with diagrams of three or even four rectangles. Suppose we have four classes A, B, C, D, and we are given that the part of A contained in B coincides with the part of C contained in D. The statement that the classes AB and CD coincide is illustrated by the shaded portion of the diagram in Figure 15.

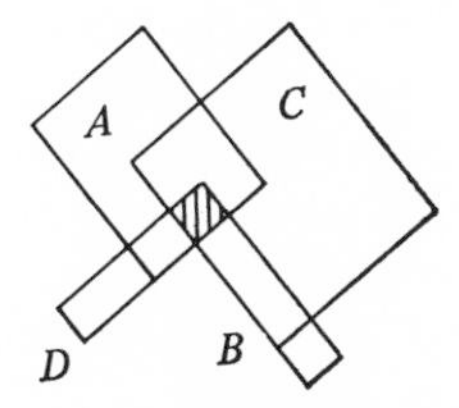

Figure 15

Some of Bolzano's contemporaries did not understand the essence of his logical insights. For example, take the following curiosity: the well-known philosophical historian F. Ueberweg considered Bolzano's *Wissenschaftslehre* to be on about the same level as *Logic for Women*, by someone named Knigge, at best a rather dubious treatise.

The first to realize the great significance of Bolzano's work in logic was the German philosopher Edmund Husserl (1859–1938). Husserl recognized that in his own *Logical Investigations* (1900–1901) he "experienced . . . a decisive influence from Bolzano"

(Reference 269, vol. 1, p. 199). For further information on the logical and philosophical contributions of Bolzano see References 265, 267, and 269–284.

2. The Problem of Quantification of the Predicate in the Work of G. Bentham and W. Hamilton. The Ideographic Devices of W. Hamilton; A Variant of the Algebraization of Syllogistics by M. W. Drobisch

In order to operate with propositions according to rules similar to those used in operating with elementary algebraic equations, one must formulate precisely the concept of the quantity of not only the subject of the statement (to which Aristotle limited himself) but of its predicate as well. Consequently, it is necessary to fix in symbolic form the precise quantitative nature of the process of predication. Ploucquet, von Holland, and other philosophers and mathematicians had already wrestled with problems similar to this one. However, only the English logicians of the early nineteenth century used the theory of quantification of the predicate for the purpose of a consistent quantitative interpretation of logical forms, thus laying the foundation for the future algebra of sets.

George Bentham and William Hamilton tried to find a transformation which would convert forms of statements containing inclusion relations between terms into equivalent forms containing relations of equality. The philosophical and general methodological outlook of most of the English logicians was free from the influence of Kant, who proclaimed that the present stagnation in logic was an inevitable result of the completeness of the Aristotelian logical analysis.

The English botanist George Bentham (1800–1884), who struggled every day with the subtleties of the operations involved in classification, gradually came to realize the necessity of their special logical analysis. Turning to the diffuse realm of logic only disillusioned Bentham: even ordinary questions of classification of the most elementary logical forms of statements were dealt with in a far from satisfactory manner.

Bentham himself decided to order simple propositions on the basis of a theory of the identity of the quantities of the subject and predicate of a statement. His conception is presented in *Outline of a New System of Logic* (London, 1827). First of all, he distinguished the following forms of identity between subjects:

"1. Between *any* individual referred to by one term and *any* individual referred to by the other. Example: the identity between equiangular and equilateral triangles.

"2. Between *any* individual referred to by one term and *any one of a part only* of the individuals referred to by the other. Example: the identity between men and animals.

"3. Between *any one of a part only* of the individuals referred to by one and *any one of a part only* of the individuals referred to by the other. Example: the identity between quadrupeds and swimming animals" (Reference 288; cited in Reference 370, p. 306).

On the basis of this classification of the types of identity relations, Bentham established his classification of the forms of simple propositions.

"Thus," he wrote, "simple propositions, considered in regard to the above relations, may therefore be either affirmative or negative; and each term may be either universal of partial. These propositions are therefore reducible to the eight following forms, in which, in order to abstract every idea not connected with the substance of each species, I have expressed the two terms by the letters X and Y, their identity by the mathematical sign $=$, diversity by the sign $\|$, universality by the words *in toto*, and partiality by the words *ex parte*; or, for the sake of still further brevity, by prefixing the letters t and p, as signs of universality and partiality. These forms are

1. X *in toto* $=$ Y *ex parte*, or $tX = pY$;
2. X *in toto* $\|$ Y *ex parte*, or $tX \| pY$;
3. X *in toto* $=$ Y *in toto*, or $tX = tY$;
4. X *in toto* $\|$ Y *in toto*, or $tX \| tY$;
5. X *ex parte* $=$ Y *ex parte*, or $pY = pY$;
6. X *ex parte* $\|$ Y *ex parte*, or $pX \| pY$;
7. X *ex parte* $=$ Y *in toto*, or $pX = tY$;
8. X *ex parte* $\|$ Y *in toto*, or $pX \| tY$"

(Reference 288).

Bentham used the symbol ‖ as an abbreviation for the expression "is not" and the symbol = for the functor "is" (in the sense of identity). Each of Aristotle's four forms of propositions, namely (symbolically) A, E, I, and O, was split by Bentham into two subforms (A_1 and A_2; E_1 and E_2, and so on), depending on whether or not the predicate in them is extensive (that is, is taken over the entire volume or quantity). Bentham's first and third forms are A_1 and A_2, respectively (here and in the following, the index 1 after a letter indicates that the predicate is not extensive in the corresponding proposition; the index 2 indicates that it is extensive); his second and fourth forms are E_1 and E_2; fifth and seventh are I_1 and I_2; sixth and eighth are O_1 and O_2.

The operation of quantification of the predicate, which allows us in many cases to make logical analysis of a proposition more precise, was, nevertheless, clearly an insufficient means for studying more subtle and special syllogisms (in particular, those for which relations are involved in the premises). In addition, as Bentham himself noted, quantification of the predicate in a negative statement would be useless from a deductive point of view.

Very little has been written about Bentham's work in logic (we can refer the reader only to Reference 287).

The logical problematics of Bentham exerted a strong influence on the work of his countryman William Hamilton. We will consider Hamilton's analyses in somewhat greater detail.

The Scottish logician and philosopher William Hamilton was born in Glasgow on March 8, 1788. His father, a doctor, occupied the chair of anatomy and botany at Glasgow University.

Initially William decided to follow in the footsteps of his father and began a strenuous study of the natural sciences, with his main effort devoted to chemistry. However, while studying at Oxford University, he unexpectedly developed a serious interest in philosophy. Gradually, his humanitarian concerns began to dominate his interests in the natural sciences, although he never completely abandoned work in the latter. Thus, Hamilton became a lawyer, in Edinburgh, and later a professor of history, philosophy, and logic.

It is interesting that Hamilton was offered the chair of logic as a result of his conflict with the Alogians and the phrenologists. By studying the skulls of men and animals, he came to conclusions which completely contradicted the contentions of the phrenologists. The results of these investigations were published in the *Edinburgh Review* in 1829 in an article entitled "Discussion on the Unconditioned." The article met with an enthusiastic response not only in England but on the Continent as well. In 1833, Hamilton also published in the *Edinburgh Review* his critique of Whately's inductive logic.

One of the sources of Hamilton's philosophical views was the concept of "healthy thought" of the Scottish philosopher Thomas Reid (1710–1796), who leaned toward the materialism and critical skepticism of Hume. However, Hamilton did not become a consistent materialist; he eclectically combined authentic materialist positions with theses that were idealist in nature (for example, the thesis that the category of causality is only an intellectual phenomenon).

Hamilton's theoretico-cognitive credo can be placed entirely within the framework of natural realism, as an attempt at philosophical comprehension of elemental materialist tendencies. Rational theology is impossible. Juxtaposition of subject and object is a necessary condition for consciousness; from this Hamilton derives the thesis of direct recognition of an object as really existing: "the only systems worthy of a philosopher are natural realism and absolute idealism . . . I believe that the external world exists because I directly know it, feel it, perceive it as existing" (cited in Reference 289, pp. 70–71).

Hamilton advanced the thesis that our knowledge is relative in that it must constantly deal with relationships. It was from a position which emphasized the importance of relativity that Hamilton criticized the idealist conceptions of Schelling and Cousin (see, for example, Hamilton's article "Philosophy of the unconditioned in reference to Cousin's doctrine of infinito-absolute," in the book *Discussions on Philosophy and Literature* (Edinburgh and London, 1866; see also Reference 291).

In 1846 Hamilton published an edition of the selected writings of his teacher Thomas Reid, with his (Hamilton's) own commentary. The beginning of his discussions with August De Morgan also dates from about that time.

Let us now proceed to an analysis of the logical views of Hamilton. The reader can find literature concerning this in References 292–294.

Logic's task, according to Hamilton, "consists in freeing the mind from errors connected with the carelessness and confusion of inconsistent thought" (Reference 290, vol. 3, p. 37). This definition stands very close to the classical Aristotelian tradition. Like Bentham, Hamilton devoted much attention to the analysis of propositions and their elementary forms. According to Hamilton, a statement expresses a comparison between concepts or subjects: "to make a statement means to *recognize relationships* [my italics — N.S.] of agreement or disagreement; these relationships contain two concepts relative to one another, two distinct subjects, or a concept and a distinct subject — for the purpose of comparing them one with another" (cited in Reference 493, p. 165).

In understanding the problem of modality of propositions, Hamilton departed from the classical tradition, treating modality as a means of expressing the degree of certainty of an assertion. According to this point of view, the term "modal" applies to all the propositions of ordinary conversational speech in which the verb is accompanied by modifiers that explain the circumstances of the action. Thus, for example, Hamilton would have called the statement "Botvinnik beat Tal" a pure (that is, nonmodal) statement and its variant "Botvinnik beat Tal with difficulty" a modal statement.

Treating the problem of quantification of the predicate in the same way that Bentham did, Hamilton, in the second volume of his lectures on logic, establishes the following eight quantified forms of statements (Reference 290, vol. 2, p. 227):

1. All A are all B.
2. All A are some B.
3. Some A are all B.
4. Some A are some B.
5. Not any A is all B.
6. Not any A is some B.
7. Some A are not any B.
8. Some A are not some B.

The reader can easily verify that these forms coincide completely with the corresponding forms of Bentham.

Hamilton called statement 1 *toto-totale* since in it the subject as well as the predicate are extensive in the sense of full quantity; for example, all equilateral triangles are (all) equiangular triangles. Statement 2 is called *toto-partiale* since in it the subject is taken extensively and the predicate nonextensively; for example, all triangles are (some) figures. Statement 3 is called *parti-totale* since in it the subject is taken nonextensively and the predicate extensively; for example, some figures are (all) triangles. Statement 4 is called *parti-partiale* since in it both the subject and the predicate are taken nonextensively; for example, some triangles are (some) equilateral figures.

Conversion of a *toto-totale* proposition gives rise to a universal proposition, whereas conversion of a *toto-partiale* proposition gives rise to a partial proposition. Conversion of a *parti-totale* proposition gives rise to a universal proposition; for example, from the truth of the statement "some geometric figures are triangles" it follows that the statement "all triangles are geometric figures" is true. Conversion of a *parti-partiale* proposition gives rise only to a partial proposition.

The considerations upon which Hamilton himself attempted to base his theory of quantification of the predicate are very interesting. He wrote that one of the cardinal errors of traditional logic was ignoring "that the predicate has always a quantity in thought, as much as the subject; although this quantity be frequently not explicitly enounced, as unnecessary in the common employment of language; for the determining notion or predicate being always thought as at least adequate to, or coextensive with, the subject or determined notion, it is seldom necessary to express this, and language tends ever to elide what may safely be omitted. But this necessity recurs the moment that, by conversion, the predicate becomes the subject of the proposition; and to omit its formal statement is to degrade logic from the science of the necessities of thought to an idle subsidiary of the ambiguities of speech. An unbiased consideration of the subject will, I am confident, convince you that this view is correct.

"1. That the predicate is as extensive as the subject is easily shown. Take the proposition 'All animal is man,' or 'All animals are men.' This we are conscious is absurd ... We feel it to be equally absurd as if we said: 'All man is all animal,' or 'All men are all animals.' Here we are aware that the subject and predicate cannot be made coextensive. If we would get rid of the absurdity, we must bring the two notions into coextension by restricting the wider. If we say, 'Man is animal' (*Homo est animal*), we think, though we do not overtly enounce it, 'All man is animal.' And what do we mean here by animal? We do not think *all*, but *some*, animal. And then we can make this indifferently either subject or predicate. We can think — we can say: 'Some animal is man,' that is, *some* is *all* man; and, *e converso*, 'Man (some or all) is animal,' viz. *some* animal ...

"2. But, in fact, ordinary language quantifies the predicate so often as this determination becomes of the smallest import. This it does either directly, by adding *all*, *some*, or their equivalent predesignations to the predicate; or it accomplishes the same end indirectly, in an exceptive or limitative form" (cited in Reference 370, pp. 307–308).

At the end of this quotation, Hamilton had in mind propositions of the type (1) only X is (is not) Y and (2) not all X having property Y are (are not) Z.

In his *Lectures on Metaphysics and Logic*, Hamilton applies an original ideographic method for illustrating logical relations and, in particular, for representing syllogistic relations.

As an early predecessor of Hamilton's ideographic devices, we should cite J. H. Alsted (1588–1638), who, in his work *Logicae systema harmonicum* (1614) already uses diagrams of linear form to give a geometric interpretation of logical formulas.

Hamilton represented the inclusion relation by wedge-shaped symbols of various lengths having their pointed ends directed either to the right or to the left (see Figure 16). The negation of the

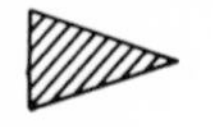

Figure 16

inclusion relation is represented by these symbols with a vertical line intersecting them. A comma, placed either before or after a letter, indicates that the corresponding term is to be taken not in full quantity; a semicolon indicates that it is to be taken in full quantity. The premises are written in one line, such that the

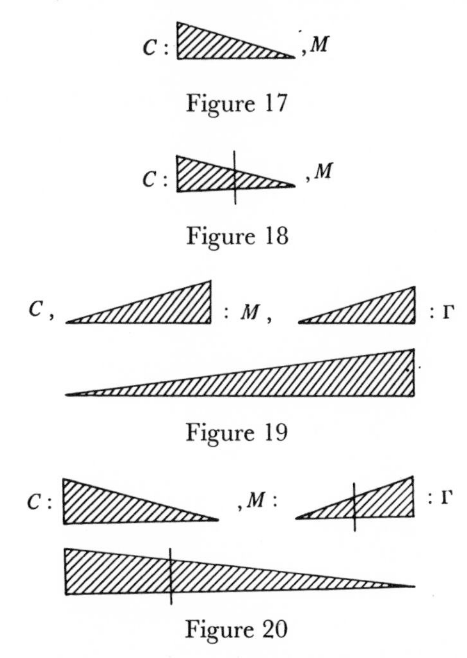

Figure 17

Figure 18

Figure 19

Figure 20

symbol M (the notation for the "mean term," repeating member, in syllogistic premises) is located in the middle. A wedge-shaped symbol joining the extreme terms indicates the conclusion.

The proposition "all C are some M" is expressed as shown in Figure 17. The proposition "not any C is some M" is shown in Figure 18.

Modus *Barbara* is given by the diagram in Figure 19, which reads: all M are some C and all Γ are some M; consequently, all Γ are some C. Modus *Cesare* is illustrated in the way shown in Figure 20 (one must read from the flat side of the wedge toward its pointed end).

Hamilton's ideographic method was an extension of the existing means of geometric illustration of quantitative logical relations.

His theory of quantification, complete, though not original, has to be considered among the stimulating prerequisites that opened the way for George Boole's logical calculus. Hamilton tried to resolve the difficulties in his theory by actually returning to the Scholastic theory of *exponibilia* (statements requiring supplementary interpretation). This latter theory, as is well-known, was connected with the peculiarities of living language and was not formally structured.

Hamilton died on May 6, 1856. In the history of British science he is also known as a passionate teacher, whose interpretation of Aristotle and the German philosophers widened the horizons of British academic education.

One of Hamilton's contemporaries on the Continent was the Leipzig philosopher M. W. Drobisch, who began to work on the formalization (more precisely, algebraization) of syllogistics even slightly before his British colleague.

Moritz Wilhelm Drobisch was born on August 16, 1802. In his philosophical views, Drobisch was a Herbartian, a pupil of the well-known German philosopher J. F. Herbert (1776–1841). The latter leaned toward so-called formalism in logic and in psychology was an advocate of the use of precise mathematical methods in dealing with the phenomena of the psychic life of man.

In 1826 Drobisch became a professor of mathematics at Leipzig, and after 1842 he was also professor of philosophy. Drobisch died in Leipzig on September 30, 1896, having discharged his professorial duties for almost seventy years.

In 1836 Drobisch's *Neue Darstellung der Logik* was published in Leipzig. This book presented Drobisch's method for the algebraization of syllogistics. In the first edition of *Darstellung* "thought" was strongly differentiated from "knowledge." In the second edition (1851), logic presented itself to Drobisch as a certain canon of thought.

For a detailed description of Drobisch's philosophical views, see References 295–297.

Drobisch presented his mathematical reworking of classical logic in the second part of the logico-mathematical appendix to *Darstellung*, where he considered the "algebraic construction of simple forms of statements and, on this basis, the reworking of syllogisms" (Reference 295, p. 131). Capital italic letters $A, B, \ldots$ are used to denote extensive (taken in full quantity) terms, lowercase italic letters $a, b, \ldots$ serve to denote nonextensive terms. Terms are considered from the point of view of the quantity (*Umfang*) of the concepts denoted by them. The symbol $<$ denotes the relation of inclusion with respect to volume or quantity (the expression to the left of the symbol is included in the expression on its right). The $=$ sign is the symbol for identity.

Drobisch also introduced an "undefined infinite concept," the analog of the present-day universal set. This concept is denoted by a capital X and is termed the *unendlicher Begriffe*. For any A and B, the following relationship holds: $(A + B) < X$, where the $+$ sign indicates the logical sum of the volumes A and B.

Drobisch used the following six forms of elementary propositions:

1. $A = b$ (literally, every A is some B).
2. $A = B$ (every A is any B).
3. $a = b$ (some A is some B or part of A is part of B).
4. $a = B$ (some A are all B).
5. $A < (X - B)$ (not a single A is B; the expression $X - B$ means for Drobisch "not-B"; in other words, negation of a concept is defined as the difference between the universal set and the positive form of the given concept).
6. $a < (X - B)$ (some A is not B).

Then Drobisch introduced five lemmas which will be used later for proving syllogistic modes as theorems. These lemmas are as follows:

1′. From $A < (X - B)$ it follows that $B < (X - A)$, and conversely.
2′. From $a < (X - B)$ it follows that $B < (X - a)$, and conversely.
3′. From $a < (A - B)$ it follows that $B < (A - a)$.
4′. $a < (A - B)$.
5′. From $A = c$ and $C < (X - B)$ it follows that $A < (X - B)$.

The operation $X - Y$, denoting the difference between the volumes of the concepts X and Y, was treated by Drobisch in the same sense as it was by Lambert. The expression $X - Y$ has meaning only if Y does not exceed X in volume. Lemma 3′ needs some explanation. In its formulation there exist two implicit conditions, which we will express explicitly: (1) $a < A$ (in the sense of the quantifiers "all" (for A) and "some" (for a); "all" includes "some"; (2) $B < A$ (by the definition of the operation of logical subtraction). The diagram in Figure 21 illustrates the validity of

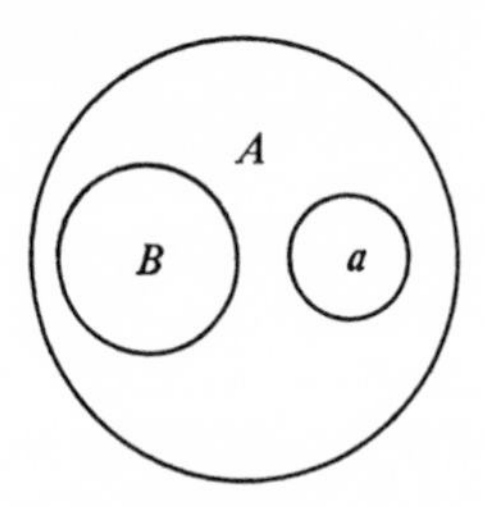

Figure 21

Lemma 3′. Indeed, if a is contained in set A and lies outside of set B, where the latter lies entirely in A, then B lies in that part of set A which is different from a.

In a similar way, we can make the formulation of Lemma 4′ more exact by saying: if $a < A$ and if $B < A$, where a does not intersect B, this implies that the relationship $a < (A - B)$ holds.

Lemma 3′ in present-day notation for set theory and propositional calculus would be represented by the relationship

$$(T_1)(B \leqslant A) \ \& \ (a \leqslant (A \cap \bar{B})) \rightarrow (B \leqslant (A \cap \bar{a})),$$

where $\leqslant$ denotes the inclusion relation, and $\cap$ denotes set intersection; in the following, $\cup$ is the symbol for union of sets. Relationship (T_1) is true. Indeed, the second premise, that is, the expression $a \leqslant (A \cap \bar{B})$, is equivalent to $(\bar{A} \cup B) \leqslant \bar{a}$. As a result of this latter relation, we obtain

$$(A \cap (\bar{A} \cup B)) \leqslant (\bar{a} \cap A) \quad \text{or} \quad (A \cap B) \leqslant (\bar{a} \cap A).$$

But according to the first premise ($B \leqslant A$), the expression $A \cap B$ coincides with B.

Therefore, the relation $(A \cap B) \leqslant (\bar{a} \cap A)$ is equivalent to the relation $B \leqslant (\bar{a} \cap A)$. This proves Lemma 3′.

Lemma 4′ in present-day notation would be represented as

$$((a \leqslant A) \,\&\, (a \cap B = 0)) \rightarrow (a \leqslant (A \cap B)).$$

We leave the proof to the reader (in this relation 0 denotes the empty set).

As for Lemma 5′, it is a somewhat more complicated form of the following well-known theorem in propositional calculus (applied here to the logic of classes): $((x \sim y) \,\&\, (y \mid z)) \rightarrow (x \mid z)$, where $\sim$ is the equivalence functor, and $\mid$ is the Sheffer stroke; the symbol $\rightarrow$ denotes "if . . . then."

In addition, Drobisch also uses the following rules: substitution of equals for equals, transitivity of the relation $<$, commutativity, and contrapositive conversion; the latter two rules correspond to the following two theorems in propositional logic:

I. $(x \,\&\, y) \rightarrow (y \,\&\, x)$,
II. $(x \rightarrow y) \rightarrow (\bar{y} \rightarrow \bar{x})$,

which can be proved without difficulty.

After formulating his lemmas, Drobisch proceeded to the algebraization of the proofs of classical syllogistic modes. Mode *Fesapo* was proved by him as follows: The major premise (not a single P is M) is written in the form

$$P < (X - M).$$

The minor premise (every M is S) is represented by the formula

$$M = s.$$

We are required to prove that there exist some S which are not P.

Proof: On the basis of Lemma 1′ and the rule about substitution of equals for equals, we transform the premise $P < (X - M)$ to the equivalent form $M < (X - P)$. Considering this latter formula together with the expression $M = s$ (the second premise), and applying to them Lemma 5′ (substituting in this lemma $A = s$,

$B = p, c = M, C = M$), we obtain the statement $s < (X - P)$, which literally means: some S are not P. Thus, we have proved the conclusion of mode *Fesapo* from its premises.

Drobisch proved the remaining Aristotelian modes in a similar manner and also proved additional modes not considered by Aristotle but obtained by means of quantification of the predicate. Like Twesten, Drobisch analyzed syllogistic chains as consisting of uniform as well as nonuniform modes. He also devoted some time to the question of topological interpretation of logical relations and, in particular, to the analysis of so-called three-circle diagrams. He used this method of illustration only for representing various combinations of sets and did not attempt to apply them to the representation of propositions, probably because he did not consider the special question of the volume (quantity) of a statement. This effort was later undertaken by the British logician John Venn (1881). Drobisch employed three-circle diagrams of the form shown in Figure 22.

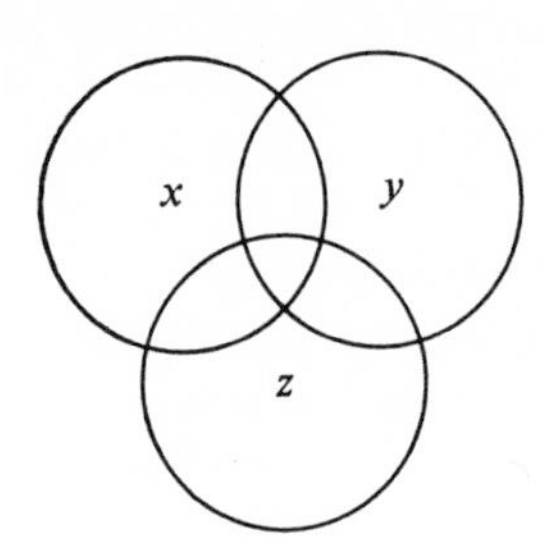

Figure 22

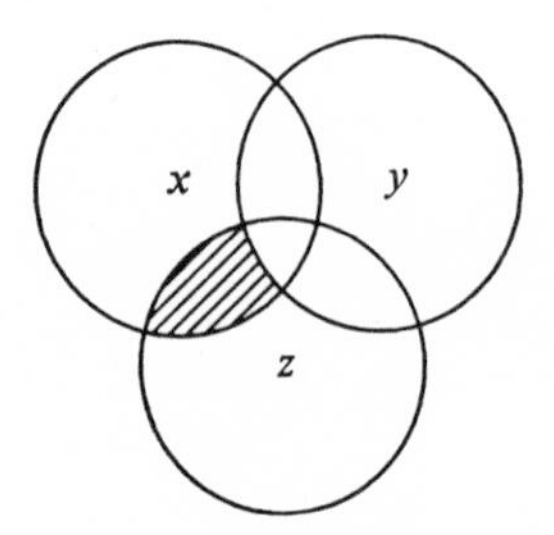

Figure 23

As an example, we will illustrate on a diagram of this type the set of those subjects x that coincide with z and are simultaneously different from y. This set is shaded on the diagram shown in Figure 23. As we know, diagrams of this type were already used by Leibniz to illustrate syllogistic modes.

Drobisch's final contribution to mathematical logic was an attempt to construct a mathematical theory of compound syllogisms. In this area, he carried on the work begun by Twesten.

3. August De Morgan, Founder of the Logical Theory of Relations

Although the beginnings of a calculus of relations can be found in the works of Lambert and Ploucquet, the English logician and mathematician De Morgan should be considered the genuine founder of the logical theory of relations. We could note in this regard the evaluation of the American mathematician and philosopher C. S. Peirce: "De Morgan was one of the best logicians that ever lived and unquestionably the father of the logic of relatives" (Reference 370, p. 434).

Of Scottish origin, August De Morgan was born in June 1806 near the city of Madras in India. He studied mathematics at Trinity College in Cambridge and in 1828 became professor of mathematics at the newly opened London University College, now a part of London University. De Morgan was a distinguished pedagogue and an ardent bibliographer; from 1847 on, he was also secretary of the Royal Astronomical Society and, in addition, the founder and first president (1866) of the London Mathematical Society.

For a long time (from 1846 to 1855) De Morgan carried on a polemic with Hamilton concerning the mathematical development of deductive logic (and, in particular, the problem of quantification of the predicate). De Morgan himself likened his discussion with Hamilton to a "fight between a cat and a dog." However, trying to be considerate of his opponent's poor health, De Morgan often softened the tone in which his polemical ardor was expressed. Hamilton was finally forced to capitulate; he then began to dispute

De Morgan's priority as an investigator of classical logical devices from a mathematical point of view. Unfortunately, both combatants were poorly informed about their predecessors (for example, Lambert and Ploucquet) and imagined themselves the first stonemasons of the foundation of symbolic logic.

De Morgan was inclined not only toward original theoretical work (for example, in the theory of mathematical series) but also toward studies in the history of mathematics and physics. After his death (De Morgan died on March 18, 1871), his book *A Budget of Paradoxes* (London, 1872) was published; it contained a wide series of articles published earlier in shorter form in the journal *Athenaeum*. The noted Russian mathematical historian V. V. Bobynin describes the nature of this work as follows:

> The problem which the author set for himself was the portrayal of all the wild theories of mathematics, physics, or astronomy which had been published since the invention of the printing press. The forms of confusion of the human mind thus presented were denoted by the author by different letters, which assigned each work to an appropriate category. Letters which appeared before the names of the various "theorists" indicated whether they were (1) anti-Copernians, (2) alchemists, (3) anti-Newtonians, (4) astrologers, (5) quadraturers of the circle, (6) mystics (primarily those concerned with the number 666), and finally (7) trisectors of angles (Reference 298, p. 832).

De Morgan was distinctly aware of the operational character of algebraic symbolism and was strongly convinced of the possibility of constructing algebras which differed from that generally used. His principal works in logic are as follows: *First Notions of Logic* (London, 1839) — here logic is defined not as the science of concepts or statements but as a theory of names of subjects; *Formal Logic or the Calculus of Inference, Necessary and Probable* (London, 1847) (this was published the same year as George Boole's *Mathematical Analysis of Logic*) — this work contains a developed system for a calculus of relations; "On the Structure of the Syllogism and its Application," *Transactions of the Cambridge Philosophical Society*, no. 1 (1849); *Syllabus of a Proposed System of Logic* (London, 1860).

Unlike Hamilton, De Morgan did not devote any detailed attention to the philosophical questions of logic. From various random notes on this subject, we can surmise his views on the relationship between names and subjects, on real definitions, and on the critique of the syllogism coming from the skeptics.

According to De Morgan, a necessary relationship between names and subjects does not exist. De Morgan mocked this concept by pointing out that believing in a necessary relationship between names and subjects is like believing that the sound of the word "man" — that is, a vibration of the air — is essentially connected to the concepts of intelligence, preparation of food by using fire, two-leggedness, etc. (Reference 305, p. 246). Later De Morgan developed a point of view that can be likened to materialism on a nominalist basis.

Criticizing the theory of definition proposed by Archbishop Whately of Dublin, De Morgan refuted the latter's method of distinction between real and nominal definitions. According to De Morgan, this question cannot be resolved by saying that a real definition always has as its subject a thing, whereas a nominal definition has as its subject a name. Rather, a real definition is that clarification of the meaning of a word (regardless of whether we have in mind the entire or only a partial meaning of the word) which is sufficient to delimit the subject named by that word from all other subjects (Reference 305, p. 36).

It is not difficult to see the similarity of this point of view to that of L. Couturat at the beginning of the twentieth century.

De Morgan refuted the reproach to the syllogism in *petitio principii* by noting that the objection to it is a result of the uselessness of the minor premise of the syllogism. Here the critics tacitly assume that one begins to consider Plato a man only after recognizing that the given individual is Plato (Reference 305, p. 259). From this it is evident that De Morgan was hardly inclined to place himself on the side of those who denied the significance of Aristotle's classical formalism.

De Morgan began his work with a generalization of the Aristotelian theory of forms of statements; here he used the sign for logical negation over not only the predicate of the statement but its subject as well (the latter operation was subsequently termed "qualification"). He noted the following eight forms of statements in syllogistics (Reference 299): SaP; $Sa\bar{P}$; SeP; $\bar{S}aP$; SiP; $\bar{S}iP$; SoP; $\bar{S}oP$, where $Sa\bar{P}$ is equivalent to SeP.

The idea of treating the negation of a concept as its complement in a "universe of arguments" (analogous to the contemporary idea of a universal set) also belongs to De Morgan. In fact, the concept of the universal set, which was to play such an important role in the logical systems of George Boole and P. S. Poretskiy, was also first suggested by De Morgan. He considered the complement of a given set to be the set of objects not contained in that given set. Both the given set and its complement were considered to lie within the limits of a certain universal set that usually could be specified (by means of extralogical criteria) in the case of a system of the concepts of some scientific discipline or subdivision thereof.

The study of conclusions drawn from quantitatively defined propositions (a subject not treated by syllogistics) also occupied an important place in De Morgan's logical investigations. An elementary example of this type of argumentation is as follows:

60 percent of S are P_1;
70 percent of S are P_2.
Therefore, at least 30 percent of S are simultaneously P_1 and P_2.

De Morgan's name is also associated with the well-known theorem in propositional calculus which states that the expression $\overline{a_1 \vee \cdots \vee a_{n-1} \vee a_n}$ is equivalent to $\bar{a}_1 \& \cdots \& \bar{a}_{n-1} \& \bar{a}_n$, and that $\overline{a_1 \& a_2 \& \cdots \& a_{n-1} \& a_n}$ is equivalent to $\bar{a}_1 \vee \bar{a}_2 \vee \cdots \bar{a}_{n-1} \vee \bar{a}_n$ (the so-called De Morgan's Laws). Here & denotes conjunction and $\vee$ denotes disjunction.

The initial step in De Morgan's reform of classical logic was his analysis of the difficulties in the treatment of copulas in statements. According to De Morgan, the copula should be considered the carrier of relations. The fact that the set of relations is infinite should not serve as a barrier to the formal analysis of the types of copulas which express them. The aim of this analysis is the discovery of general copular conditions, such as symmetry and transitivity. To carry out this analysis, De Morgan had to construct a developed symbolic language, distinguishing objects (denoted by the letters $X, Y, Z, \ldots$) and relations (denoted by the letters $L, M, N, \ldots$) and also including symbols for logical functors (which differ from one work to the next; we will adopt the notation of De Morgan's last investigations).

De Morgan chose not to consider as the elementary type of statements those reducing to the Aristotelian representation "X is (is not) Y"; instead he considered that an elementary statement has the structure $X..LY$, denoting "that X is some one of the objects of thought which stand to Y in the relation L or is one of the L's of Y" (cited in Reference 370, p. 434). In De Morgan's notation $X.LY$ signifies that X is *not* any one of the L's of Y.

Translating the affirmative expression $(X..LY)$ into present-day language, we can write it in the form xLy, which, according to De Morgan, is equivalent to the expression $x = L(y)$, where x is the "subject" and y is the "predicate."

De Morgan's algebra of relations included the following six fundamental operations:

1. MN' (the logical sum of the relations M and N);
2. MN (the logical product of these relations);
3. n (the operation of obtaining the complementary relation to N; literally, not-N; De Morgan called n the contrary relation);
4. N^{-1} (the operation of obtaining the converse relation or the converse of N);
5. $M \between N$ (the operation of generating the relative sum of the relations M and N);
6. $M(N)$ (the operation of generating the relative product of the relations M and N or the "composition" of those relations).

Let us now give the contemporary transcription of the above operations, at the same time verifying them by using texts taken directly from the work of De Morgan. Transcribing Operations 1 through 6, we obtain

1′. $x(M \cup N)y$, which is equivalent to $xMy \vee xNy$;
2′. $x(M \cap N)y$, which is equivalent to $xMy \,\&\, xNy$;
3′. $x\bar{N}y$, which is equivalent to $\overline{xNy}$;

4′. $x\breve{N}y$, which is equivalent to yNx;
5′. $x(M * N)y$, which implies $\forall z(xMz \vee zNy)$;
6′. $x(M/N)y$, which implies $\exists z(xMz \mathbin{\&} zNy)$.

Here $\cup$ denotes the operation of union of relations, $\cap$ denotes their intersection, $\forall$ is the universal quantifier, $\exists$ is the existential quantifier, $*$ denotes the relative sum, / denotes the composition of relations, & denotes conjunction, and $\vee$ denotes disjunction.

Here is the text where De Morgan introduces the expressions and the relations necessary in the theory of syllogisms "We have thus three symbols of compound relation: LM, an L of an M; LM', an L of every M; L, M, an L of none but M's. No other compounds will be needed in syllogism, until the premises themselves contain compound relations" (cited in Reference 370, p. 435). The passage which introduces the operations of conversion and complementation of relations is:

"The converse relation of L, L^{-1}, is defined as usual: if $X..LY$, $Y..L^{-1}X$; if X be one of the L's of Y, Y is one of the L^{-1}'s of X. And L^{-1} may be read 'L-verse of X.' Those who dislike the mathematical symbol L^{-1} might write L^{v}. This language would be very convenient in mathematics: $\varphi^{-1}x$ might be the 'φ-verse of x' read as 'φ-verse x.'

"Relations are assumed to exist between any two terms whatsoever. If X be not any L of Y, X is to Y in some not-L relation: let this *contrary* relation be signified by l; thus, $X.LY$ gives and is given by $X..lY$. Contrary relations may be compounded, though contrary terms cannot: Xx, both X and not-X, is impossible; but Llx, the L of a not-L of X, is conceivable. Thus a man may be the partisan of a nonpartisan of X" (Reference 370, p. 436). And, finally, De Morgan defines the concept of "composition" of relations:

"When the predicate is itself the subject of a relation, there may be a composition: thus if $X..L(MY)$, if X be one of the L's of one of the M's of Y, we may think of X as an 'L of M' of Y, expressed by $X..(LM)Y$, or simply by $X..LMY$" (Reference 370).

De Morgan wanted to say that in the representation of L/M by $\exists z(xLz \mathbin{\&} zMy)$, the term z is the predicate in the statement xLz

and the subject in the statement zMy (the predicate of one relation is itself the subject of another relation). In a similar way (by taking into account the duality relationship, which De Morgan understood very well but did not explicitly define), we can define the concept of a relative sum $L * M$.

Having established the fundamental operations of his logic of relations, De Morgan then went on to formulate a number of theorems in this algebra. "Contraries of converses are converses: thus not-L and not-L^{-1} are converses. For $X..LY$ and $Y..L^{-1}X$ are identical; whence $X..$not-LY and $Y..($not-$L^{-1})X$, their simple denials, are identical; whence not-L and not-L^{-1} are converses.

"Converses of contraries are contraries: thus L^{-1} and (not-$L)^{-1}$ are contraries. For since $X..LY$ and $X..$not-LY are simple denials of each other, so are their converses $Y..L^{-1}X$ and $Y..($not-$L)^{-1}X$; whence L^{-1} and (not-$L)^{-1}$ are contraries.

"The contrary of a converse is the converse of the contrary: not-L^{-1} is (not-$L)^{-1}$. For $X..LY$ is identical to $Y.$not-$L^{-1}X$ and to $X.$ (not-$L)Y$, which is also identical to $Y.$ (not-$L)^{-1}X$." (Cited in Reference 370, p. 437.)

Thus, De Morgan asserts that (1) negation of mutually converse relations gives mutually converse relations, (2) conversion of mutually complementary relations gives mutually complementary relations, and (3) the complement of the converse is the converse of the complement.

It will be easiest to prove the third statement first, as a lemma; then this lemma can be used to prove the first two. Indeed, $\overline{x\breve{L}y}$ can be replaced by $\overline{yLx}$ (by the definition of a converse relation), which is equivalent to $y\bar{L}x$ (transfer of the negation from the proposition as a whole to its copula); from this, again by the definition of the converse, we obtain $x\breve{\bar{L}}y$. Thus, we have proved that $x\bar{\breve{L}}y$ is equivalent to $x\breve{\bar{L}}y$.

Now let us prove the first theorem. Suppose we are given two relations L_1 and L_2 such that $xL_1y = xLy$ and $xL_2y = x\breve{L}y$. We must prove that $\bar{L}_1$ and $\bar{L}_2$ are mutually converse.

Forming the complement of L_1, we have $x\bar{L}y$. Taking the complement of L_2, we obtain $x\bar{\breve{L}}y$, which, by the preceding lemma, is equivalent to $x\breve{\bar{L}}y$. But $x\bar{L}y$ and $x\breve{\bar{L}}y$ are mutually converse, which proves the theorem.

Now we consider the second theorem. We are given two relations L_1 and L_2 such that $L_1 = xLy$ and $L_2 = x\bar{L}y$. We must show that $\breve{L}_1$ and $\breve{L}_2$ are mutually contrary, that is, they exclude one another. We form the converse of L_1, which gives $x\breve{L}y$, and the converse of L_2, which gives $x\breve{\bar{L}}y$. The expression $x\breve{\bar{L}}y$, by the preceding lemma, is equivalent to $x\breve{\bar{L}}y$, which proves the mutual contrariness of the relations $\breve{L}_1$ and $\breve{L}_2$.

De Morgan went on to prove many theorems about converse and complementary relations, establishing, for example, that if xL_1y is contained in xL_2y, then $x\breve{L}_1y$ is also contained in $x\breve{L}_2y$ and, at the same time, $x\bar{L}_2y$ is contained in $x\bar{L}_1y$: "If a first relation be contained in a second, then the converse of the first is contained in the converse of the second: but the contrary of the second is contained in the contrary of the first" (cited in Reference 370, p. 437). And further on: "The conversion of a compound relation converts both components and inverts their order" (*ibid.*). For example, we have the equation

$$x\overparen{(L \cap M)}y = y(\breve{L} \cap \breve{M})x.$$

De Morgan divided relations into two main classes: transitive and nontransitive.

Having given a completely rigorous definition of the transitivity of a relation, De Morgan went on to conclude: "A transitive relation has a transitive converse, but not necessarily a transitive contrary" (Reference 370, p. 437). In fact, if, for example, the relation "greater than" is transitive, then the relation "less than" is also transitive; however, from the premises "x is not the ancestor of y" and "y is not the ancestor of z," it does not follow that "x is not the ancestor of z."

De Morgan was the initiator of the application of logical calculuses to the verification of theorems in the theory of probability, anticipating the analogous efforts of George Boole. By doing this, he hoped to demonstrate the practical usefulness of his calculus;

thus, he translated the concepts into the language of the theory of probability. For example, the concept of the relative product of relations was made parallel to the concept of conditional probability.

De Morgan's logical work was not sufficiently well understood by his contemporaries. The complicated and not always uniform notation frightened many readers, who were pragmatically inclined to demand of any new theory immediate practical applications. The fundamental historical significance of De Morgan's system is principally that it stimulated the development of the algebra of relations of C. S. Peirce and gave a push to George Boole in his attempts to generalize Aristotelian syllogistics by using the calculus of classes, which had clearly not been completely developed by De Morgan.

Literature on the logic of August De Morgan can be found in References 298–306.

CHAPTER FIVE

George Boole's Calculus of Classes

1. A Brief Biography of George Boole

George Boole came from a family of petty tradesmen who lived in the vicinity of the city of Lincoln. He was born on November 2, 1815. George's father, John Boole, was a bootmaker, but his real interests were directed toward mathematics. He was also attracted by the manufacture of various optical instruments. The inhabitants of Lincoln knew John Boole well: he was a zealous advocate of earlier prescription and wearing of glasses, and with his own hands he built a telescope which was not at all bad for those times.

George received a very meager education. After completing elementary school, he attended a commercial college for a short time. But commerce held no charm for the youth, and he soon decided to leave. At the same time, there matured in him a strong desire to become a well-educated man. He began to take lessons in Latin from the bookseller William Brooke, who became his close friend.

The inquisitive young man taught himself Greek, and later German and French from books he borrowed from his friend. At the age of fourteen, Boole completed a verse translation of the ancient Greek poet Meleager's *Ode to Spring*. This translation was printed in Lincoln and thus actually became Boole's first published work.

At sixteen George Boole became an assistant teacher at a private school in Duncaster, at the same time fulfilling the duties of a laboratory assistant and a porter. Not wishing to lose any time, the seventeen-year-old laboratory assistant began a systematic study of mathematics. According to Mrs. Boole (Reference 347), her husband later told her that he began to read mathematical books because they were much cheaper than books in classical philology.

In his first attempts at mathematical self-education, Boole received help from his father and also from his friend D. S. Dixon, who had received a mathematics degree from Oxford.

In 1833 Boole left the school at Duncaster and went to a school in the Lincoln suburb of Waddington. There he became the head teacher. In 1840 Boole founded his own school in Lincoln.

The range of Boole's scientific interests was rather wide: he read with equal zest mathematics, logic, the ethics of Spinoza, and the philosophical works of Aristotle and Cicero. But gradually he began to lean more and more toward problems involving the application of mathematical methods to the realm of the humanities (which at that time was thought to include the field of logic). Boole studied Newton's *Principia* and Lagrange's *Mechanics* very carefully, comparing the methods of both scientists as he went along. He was struck by Lagrange's ability to reduce the expression of physical problems to purely mathematical terms. Evidently, Boole was already seriously reflecting on the possibility of abstracting from physical facts and the data of ordinary conversational speech and making a transformation to some system of effectively constructed symbols; such a system would have to be operable according to certain inherent laws and would have to be self-sufficient.

Boole began a correspondence with some mathematicians from Cambridge, who noted the originality of his mathematical ideas and advised him not to hide them away. Taking the advice of his new friends, Boole began to publicize some of his work and in 1844 received a gold medal for his research in mathematical analysis. In 1849 Boole's mathematician friends helped him secure a position as Professor of Mathematics in the newly established Queen's College in Cork, Ireland. Boole was accepted for this position in spite of the fact that he did not have a university education or a degree.

In his mathematical work, Boole devoted persistent attention to "symbolical reasoning." The English logician felt that mathematical operations (including even those like differentiation and integration) should be studied, first of all, from the point of view of their formal properties. This would make it possible to carry out various transformations on expressions containing these operations, independently of the internal content of such expressions.

R. G. Rhees (Reference 342) gives us a picture of Boole as a teacher. He quotes a recollection of Boole by one of his students, R. Jameson, who went to work as a teacher in Shanghai. Jameson recalls that Boole usually tried to teach in such a way that his listeners would be able to uncover for themselves some of the scientific results already obtained by others (that is, he didn't present them all in his lectures). He taught us, continues Jameson, to feel "the joy of discovery" (Reference 342). Jameson and Rhees could have added that Boole obviously maintained the hope that some day his students would also make "undiscovered" discoveries.

Boole's concern with logic was to a significant degree stimulated by the discussion between De Morgan and Hamilton, which he followed with great interest during the spring of 1847. He notes this fact in the preface to *The Mathematical Analysis of Logic*, which he wrote in October, 1847. Boole recognized, as well, that De Morgan was the first logician to turn to the analysis of "numerically definite" propositions.

Along with his logical and mathematical investigations, Boole continued to write poetic works which were classical in form and philosophical in content.

In 1855 Boole married Mary Everest, the daughter of Thomas Everest and the niece of George Everest, a professor of Greek at Queen's College. Boole had five daughters: Mary, Lucy, Alice, Margaret, and Ethel. Lucy Boole became the first woman in England ever to receive the title of Professor of Chemistry. Alice, who, like her father, did not receive any special mathematical education, nevertheless made a number of mathematical discoveries. In particular, she constructed from cardboard, by a purely Euclidean method using only a ruler and a compass, three-dimensional sections of all six regular four-dimensional figures. Boole's youngest daughter became a writer and was well-known as the author of a popular novel about the battle for freedom of the Italian carabinieri.

Boole's contemporaries took note of his democratic attitudes and the complete absence in him of any regard for the social prejudices and barriers that existed in Britain. He was an extremely principled man and also had a very good sense of humor.

Boole died on December 8, 1864 in Bellintemple, near Cork. Of those of Boole's works published during his lifetime, we note the following: *The Mathematical Analysis of Logic* (Cambridge); "The Calculus of Logic," *The Cambridge and Dublin Mathematical Journal* (May, 1848); *An Investigation of the Laws of Thought* (London, 1854); *A Treatise of Differential Equations* (Cambridge, 1859), and *A Treatise of the Calculus of Finite Differences* (Cambridge, 1860). For a short bibliographical list of editions of Boole's work and books or articles concerning him, see References 307–370.

Boole died ten years after his basic logical work (*Investigation of the Laws of Thought*) was published. However, the manuscripts he left indicate that he intended to continue his development of logical theory (Reference 309). Beginning in 1854, Boole concentrated his efforts on the application of his calculus to the theory of probability; during the next ten years he did not publish any works directly related to logic. However, Boole's work in mathematics was always a help and a stimulation to his reflections on logic, even when, in the last period of his creative efforts, he was coming to the conclusion that logic is independent of mathematics and should form its basis (Reference 309, p. 11).

Boole proposed the following interpretation of the calculus of classes to fit the needs of the theory of probability:

1. the probability of the class 1 (the universal class) is equal to 1;
2. the probability of the complement of x, the class $1 - x$, is equal to the probability of class 1 minus the probability of class x;
3. the probability of the class $x + y$ is the sum of the probabilities of the classes x and y;
4. the probability of the class xy, where x and y are independent, is the product of the probabilities of x and y.

This interpretation is introduced in an article which was not published during Boole's lifetime: "On Propositions Numerically Definite," which was written after the publication of De Morgan's *Formal Logic* but before the completion of *Investigation of the Laws of*

Thought (1854). The article was later published at the insistence of De Morgan in *Transactions of the Cambridge Philosophical Society*, vol. XI, pt. II (Cambridge, 1868).

2. Boole's Methodological Ideas

The question of Boole's methodological views is a rather complicated one. In most cases, these must be judged on the basis of his early statements only. Their analysis usually leads to the conclusion that Boole can be considered a forerunner of formalism of the type associated with Hilbert (H. Scholz, Reference 369) with elements of psychologism in the spirit of John Stuart Mill (J. Jorgensen, Reference 326, pp. 97–99). Scholz, in particular, noted that Boole anticipated Hilbert's idea of a mathematical theory and the related concept of mathematics as a purely structural discipline. In fact, he did this with such preciseness that Hilbert could scarcely have improved upon it (Reference 310, p. 8).

As a justification of his point of view, Scholz cited, in particular, the following statement by Boole:

"They who are acquainted with the present state of the theory of Symbolical Algebra are aware that the validity of the processes of analysis does not depend upon the interpretation of the symbols which are employed, but solely upon the laws of their combination. Every system of interpretation which does not affect the truth of the relations supposed is equally admissible, and it is thus that the same process may under one scheme of interpretation represent the solution of a question on the properties of numbers, under another that of a geometrical problem, and under a third that of a problem of dynamics or optics." (Reference 310, pp. 8–9).

Jorgensen noted that, in the debate between the psychologists and the antipsychologists on the interpretation of the nature of logical relations, Boole unreservedly stood on the side of the former (Reference 326, p. 98).

However, we cannot agree with the treatment given Boole's methodological principles by Scholz and Jorgensen. First of all, both authors appeal only to early statements by Boole, ignoring the evolution of his general views. In addition, it seems to us that even these early statements by Boole do not justify such unqualified characterizations. We must return to the views of Boole himself. He stated that the aim of his *Laws of Thought* is the study of the fundamental laws of those operations of the mind by which reasoning is carried out — in order to give expression to those laws in the symbolic language of logical calculus and on that basis to verify that logic is a science and that its methods are valid. Furthermore, Boole sought to make these methods the basis of a more general method, for the purpose of applying it to the mathematical theory of probability; and, finally, he hoped, by uniting various elements of truth, to continue the advancement toward certain probable indications concerning the nature and structure of human thought (Boole, Reference 327, pp. 1–2).

Boole was a complete stranger to subjectivism; he recognized the existence of laws of thought as objective compulsory relations, independent of the will of the apprehending individual. He asserted that the laws of reasoning have objective existence as laws of human thought (Reference 327, pp. 39–40). Furthermore, he felt that his views on the science of logic should have a material source (Reference 327, p. 22).

Here there is no trace of Mill's psychologism and subjectivism. The desire to penetrate into the actual laws governing cognitive thought can hardly be classified as psychologism. Obviously Jorgensen mistakenly applied to Boole's natural materialist purposes the odious term "psychologism."

According to Boole, general studies of logic should also throw light on the nature of intellectual capabilities. From this, one can only conclude that he did not ignore the practical aspect of logical investigations. In this respect, the disclosure of the nature of syllogistic reasoning was of special significance from Boole's point of view. Boole felt that the formulation of logic as a calculus was hardly an arbitrary act but one dictated by the identical formal characteristics possessed by logical transformations (on the one hand) and mathematical operations (on the other). Boole expressed the opinion that the subject of "mathematics is limited only to the concepts of number and quantity" (Reference 327, p. 12), which would appear to us to be incorrect. A few lines later, however,

Boole categorically declares: "concern with the ideas of number and quantity does not comprise the essence of mathematics."*

Nor were Boole's antiformalistic views a secret to his colleagues. One of Boole's contemporaries and a reviewer of his work, R. Harley, remarked that Boole's system of the basic laws of thought is not derived from the theory of numbers, as is usually stated, but from an analysis of the object itself (Reference 311, p. 3). He went on to say that Boole's calculus hardly consists in the reduction of the concepts of logic to the concept of quantity (Reference 311, pp. 3–4).

Boole drew a distinction between ordinary conversational language and symbolic "language," which he considered a set of defined symbolic devices and the methods of operating with them, as is found in the so-called exact sciences. *Both* are "instruments of human reasoning," differing only in the degree of accuracy and definiteness. Boole was deeply convinced that, by studying the laws of signs (symbolic language), one could in reality come to know the fundamental laws of reason (Reference 327, p. 24). From his point of view, the laws of thought correspond to (or, better, are reflected by) the laws of operating with logical symbols, the latter being introduced only as auxiliary devices for the study of actual thought and its laws.

In concluding this section, we should note one more very interesting fact, namely, the attempt by one of Boole's contemporaries to apply the Boolean system to chemistry. Reference 312 takes note of an article by Professor P. D. Tates which reviews the work of one Benjamin Brody, who attempts to reinterpret Boolean formalism for the purpose of proving certain points in chemical theory. According to Tates, Brody attempted to do in chemistry what Boole did in logic, that is, reduce chemistry to a subdivision of mathematics (using this term in a somewhat wider sense than is usual) (Reference 312, p. 15). Brody himself pursued the goal of freeing the science of chemistry from errors created by speculative hypotheses and of arming it with the necessary instrument for progressive thought and exact language (Reference 312, p. 15).

* What Bertrand Russell said about Boole is of some interest here. "Pure mathematics," wrote Russell, "was discovered by Boole" (Reference 344).

Harley justifiably noted that Benjamin Brody's system was obviously stimulated by Boole's *Investigation of the Laws of Thought.*

3. Basic Characteristics of Boole's Logical System

From the beginning, we will attempt to present Boolean calculus as it was first formulated by Boole himself. In such a presentation, there may be much that will seem strange and artificial to the reader. The first surprising thing is the ease with which Boole transferred to logic the laws of arithmetic and the rules for arithmetic operations. This even amazed his own contemporaries. To understand Boole's train of thought, we will adhere very closely to his original presentation; only later will we attempt to translate it into a language that is closer to the usage in present-day mathematical logic. The basic operations in Boole's logic are as follows:

1. Addition, denoted by the symbol $+$. In the calculus of classes, Boole's formula $x + y$ corresponds to the union of the classes x and y minus their common portion; in the propositional calculus it is so-called "strict disjunction." In terms of union (corresponding to disjunction) $\cup$, intersection (corresponding to conjunction) $\cdot$, and complementation (corresponding to negation) $-$, Boolean addition can be expressed as follows:

$$x + y = x \cdot \bar{y} \cup \bar{x} \cdot y.$$

The inclusive "or" can be expressed in terms of Boolean addition by use of the two formulas $x \vee y = x + y + xy$ and $x \vee y = x + \bar{x}y$.

Let us consider for a moment the remarks of the well-known logician and historian of logic C. I. Lewis (Reference 364) on how the prominent American logician and mathematician C. S. Peirce utilized the system of operations in Boole's logic. According to Lewis, Peirce used the following logical operations in Boolean calculus:

a. $a \,+\!\!\!\!\!,\; b$ denotes the class of those things which are either a or b or simultaneously a and b.

b. The inverse of the preceding operation is the operation $a \vdash b$, such that, if $x \,+\!\!\!\!\!,\; b = a$, then $x = a \vdash b$.

c. ab denotes the class of those things which are simultaneously a and b.

d. The inverse of the preceding operation is the operation a/b, such that, if $bx = a$, then $x = a/b$. *This is the Boolean operation* a/b [my italics — N.S.].

The arithmetic operations were the following:

e. $a + b$ denotes the class of those things which are either a or b but not both simultaneously. *This is the Boolean sum* $a + b$ [my italics — N.S.].

f. The inverse of the preceding operation is the operation $a - b$, denoting the class of those a which are different from b.

g. $a \times b$ and $a \div b$ are the exact analogs of the corresponding arithmetic operations. They do not have any relationship to corresponding logical operations, as do the operations $a + b$ and $a - b$. Peirce does not use them explicitly in applying his system to the theory of probability (Reference 364, pp. 81–82, as cited in Reference 326, pp. 129–130).

Thus, from this interesting report, we can see that Peirce considered ordinary disjunction to be logical, and strict Boolean disjunction to be an arithmetic operation. From here it is not very far to the idea that the algebra of logic can be construed as a ring of residues modulo two (see, for example, Reference 9). Let us now return to consideration of the remaining operations in Boole's logic.

2. Multiplication is denoted by the symbol $\cdot$ or simply by one expression written next to the other (without any intervening symbol); in the calculus of classes this operation corresponds to intersection, in the propositional calculus it corresponds to conjunction.

3. Complementation of x with respect to 1 is denoted by the expression $1 - x$; in the calculus of classes the formula $1 - x$ indicates the complement of the class x; in the propositional calculus it denotes the negation of x.*

The fundamental relation used in Boole's logical system is the relation of equality: by means of this relation, other relations are

* Boole also considers subtraction, which he understands as the operation inverse to the operation of addition. He defines subtraction as $x - y_{Df} \equiv x(1 - y)$.

defined. The first relation defined by use of it is set inclusion. Boole used several definitions of the inclusion relation. One of them says $x \leqslant y_{Df} \equiv x(1 - y) = 0$, where 0 is the symbol for the null (empty) class.

Boole elucidated the properties of the operations he has introduced, establishing the following laws:

The commutative law for multiplication: $xy = yx$;
The associative law for multiplication: $x(yz) = (xy)z$;
The distributive law for multiplication with respect to addition:

$$z(x + y) = zx + zy;$$

The distributive law for multiplication with respect to subtraction:

$$z(x - y) = zx - zy.$$

The following are considered fundamental properties of the Boolean relation of equality:* (m) if $x = y$, then $zx = zy$ and (n) if $x = y$, then $z + x = z + y$.

The essential difference between logic and algebra, according to Boole, consists in the fact that in the former the law $xx = x$ (the principle of idempotence) always holds, whereas in the latter the equation $x^2 = x$ holds only if $x = 1$ or $x = 0$, the two roots of the equation $x^2 - x = 0$.

Continuing, Boole asserted that, if we imagine an algebra in which the symbols admit the values 0 and 1, and no other values, then the laws, axioms, and transformations of this algebra will be completely identical to the laws, axioms and transformations of the algebra of logic (Reference 327, pp. 37–38).

Having assumed that the fundamental objects, relations, and operations of logic can be expressed by means of symbols satisfying the laws of an elementary algebra for 0 and 1, Boole concluded that, in carrying out the individual steps of a calculational procedure in the algebra of logic, one can disregard the direct logical

* We do not consider it our task here to axiomatize Boole's system, just as we do not attempt to justify the rules presented. We wish only to give a general outline of Boole's calculus. An analysis of Boole's procedures is given in Section 5 of this chapter.

interpretation of the symbols. Thus, one can calculate with these symbols, considering them as quantitative signs, exactly as with ordinary algebraic rules for 0 and 1. After carrying out the required processes for solution, one can restore the actual logical meanings in the final result (Reference 327, p. 70).

This is the fundamental principle of the logical calculus developed by Boole; it expresses the formal agreement existing between quantitative algebra and the algebra of logic, regardless of the difference between the objects of investigation in these two disciplines.

Boole devoted special attention to the study of logical functions and operations with them. He did not give any special definition of a logical function, remarking only that in expressions of the form $f(x), f(xy), \ldots$ the symbols $x, y, \ldots$ could be considered as logical symbols, admitting only the values 0 and 1. Next, Boole defined the concept of the expansion of a function. Every function $f(x)$ where x is a logical symbol (or a quantity admitting only the values 0 and 1) is called expanded if it is reduced to the form $ax + b(1 - x)$, where a and b are determined so that this form is equivalent to the original function (Reference 327, p. 72).

Boole derived the general formula for expansion of a logical function as follows. Setting $f(x) = ax + b(1 - x)$, he first took $x = 1$, obtaining $f(1) = a \cdot 1 + b(1 - 1) = a$, and thus $a = f(1)$; then he set $x = 0$, which gives $f(0) = a \cdot 0 + b(1 - 0) = b$. From these the equation

$$f(x) = f(1)x + f(0) \cdot (1 - x).$$

follows. Generalizing the preceding equation, Boole obtained

$$f(x, y) = f(1, 1)xy + f(1, 0)x(1 - y) + f(0, 1)(1 - x)y + f(0, 0)(1 - x)(1 - y)$$

and, continuing,

$$f(x, y, z, \ldots) = f(1, 1, 1, \ldots)xyz \ldots + f(1, 0, 1 \ldots, x)(1 - y)z \ldots + f(1, 1, 0 \ldots)xy(1 - z) \ldots + \cdots + f(0, 0, 0, \ldots)(1 - x)(1 - y)(1 - z) \ldots.$$

Boole called the terms containing the symbols x, y, z the constituents of the expansion; terms of the form $f(0)$, $f(1, 1, \ldots)$, $f(0, 0, \ldots), \ldots$ were called the coefficients of the constituents. Then he sought to strengthen the expansion he had introduced by drawing an analogy to the Maclaurin expansion of differential calculus (Reference 307, pp. 78–79). The Maclaurin series has the form

$$f(x) = f(0) + f'(0)x + f''(0)\frac{x^2}{1\cdot 2} + \cdots. \tag{1}$$

Now we apply the principle of idempotence, according to which $x^n = x$. Then Formula 1 can be reduced to the form

$$f(x) = f(0) + \left(f'(0)x + \frac{f''(0)}{1\cdot 2}x + \frac{f'''(0)}{1\cdot 2\cdot 3}x + \cdots\right). \tag{2}$$

Setting $x = 1$, we obtain

$$f(1) = f(0) + f'(0) + \frac{f''(0)}{1\cdot 2} + \frac{f'''(0)}{1\cdot 2\cdot 3} + \cdots. \tag{3}$$

Hence, it follows that

$$f'(0) + \frac{f''(0)}{1\cdot 2} + \cdots = f(1) - f(0). \tag{4}$$

Then, taking into account Formula 2, we have

$$f(x) = f(0) + (f(1) - f(0))x, \tag{5}$$

which is equivalent to*

$$f(x) = f(1)x + f(0)(1 - x). \tag{6}$$

Boole's calculus of classes actually presupposes a system of concepts from ordinary substantive logic, for example, logical inference, deduction, and so on. Treating algebra as a science of equalities, he also wrote logical expressions in the calculus of classes in the form of equalities. Boole attempted to give complete explicit definitions of the form $x = f(t)$, from which it would be possible to define all the properties of the specified object. Speaking in contemporary terms, this would correspond to an attempt to replace axiomatic definitions with explicit ones.

* Boole's method of drawing an analogy to the Maclaurin series is at best only a heuristic device; it is hardly a rigorous proof. In carrying out this analogy, Boole himself did not point out all of the assumptions he made; among them is the extension of the distributive law to the case of an infinite series.

By analogy with arithmetic algebra, Boole also posed the problem of excluding certain terms from given logical equations (the problem of elimination). The essence of this problem is as follows: suppose we are given a certain relationship containing several classes $x, y, z, \ldots$. We wish to derive from this relationship a new relationship, equivalent to the conjunction of all of those corollaries of the given relationship which contain only certain specified ones of the classes $x, y, z, \ldots$; that is, we wish to establish what can be said about the relationship between only certain specified classes on the basis of a statement that contains these specified classes and also other additional classes.

To eliminate the letter x from a logical equation of the form $f(x) = 0$, Boole suggested the following method. First we expand the expression $f(x)$ into constituents: $f(1)\cdot x + f(0)\cdot(1 - x) = 0$. Then, solving this expression algebraically with respect to x, we obtain

$$x = -\frac{f(0)}{f(1) - f(0)} \quad \text{or} \quad x = \frac{f(0)}{f(0) - f(1)}. \tag{I}$$

This yields

$$1 - x = 1 - \frac{f(0)}{f(0) - f(1)}. \tag{II}$$

Substituting Expressions I and II into the formula for the "fundamental law" $x^2 = x$ or $x(1 - x) = 0$, we obtain

$$-\frac{f(0)\cdot f(1)}{(f(0) - f(1))^2} = 0,$$

and, hence, $f(1)\cdot f(0) = 0$.*

* We note that in the formula

$$\frac{f(0)\cdot f(1)}{(f(0) - f(1))^2}$$

Boole obviously assumed that $f(0) \neq f(1)$. Boole's theorem for elimination can easily be generalized to the case of more than one variable. Let us exclude, for example, x and z from the equation $f(x, z) = 0$. First eliminating x, we obtain $(1, z)\cdot f(0, z) = 0$. Then, eliminating z, we obtain

$$f(1, 1)\cdot f(0, 1)\cdot f(1, 0)\cdot f(0, 0) = 0. \tag{III}$$

In general, the number of multipliers in formulas of type III (i.e., in the results of an elimination) is 2^m, where m is the number of variables being eliminated.

Boole formulated his rule for elimination by stating that, if $f(x) = 0$ is an arbitrary true logical equation containing the symbol for the class x, either with or without symbols representing other classes, then the equation $f(1) \cdot f(0) = 0$ will be true, independent of the interpretation of x, since it is an exact result of elimination of x from the original equality. In other words, elimination of x from a given equation $f(x) = 0$ can be accomplished by successive substitution in this equation of $x = 1$ and $x = 0$, followed by multiplication of the two equations obtained through these substitutions (Reference 327, p. 101).

Let us now describe the Boolean methods of solution of logical equations. These methods, according to Boole, correspond to a generalized representation of classical logical concepts. He noted that the most general problem of logic can be formulated as follows: given a logical equation containing the symbols $x, y, z, w; \ldots$, find a logically interpretable expression for explaining the relationship of the class denoted by w to the classes denoted by x, y, z (Reference 327, p. 87).

What Boole has in mind here is obvious from the methods of solution he proposed for this problem.* In its general form, the process of solution can be formulated as follows: we can find the required expression for w if we first expand the given equation with respect to w, obtaining an equation of the form

$$Ew + E'(1 - w) = 0, \tag{1}$$

where E and E' are functions of the symbols $x, y, z, \ldots$ Then, using the rules of arithmetic algebra, we solve Equation 1 for w. We obtain $E' = (E' - E)w$ and, thus,

$$w = \frac{E'}{E' - E}. \tag{2}$$

The right-hand side of Equation 2 can then be fully expanded with

* Elsewhere, Boole formulated the same problem more precisely as: "Given the equation, to express one term as a function of the others." (Cited in Reference 340, p. 99). In Boole's opinion, this is one of the fundamental problems of logical calculus.

respect to the terms $x, y, z, \ldots$ In this case, each of the coefficients of the constituents of this expansion may take any one of the following set of possible values:

$$\frac{1}{1}; \frac{0}{1}; \frac{1}{0}; \frac{0}{0}; \frac{m}{1}; \frac{1}{n}; \frac{m}{0}; \frac{0}{n}; \frac{m}{n};$$

where m and n are whole positive or negative numbers, different from 1.

Then Boole proposed a series of rules for interpreting the coefficients of the constituents in these types of expansions, to transform them into purely logical functions. These rules of interpretation are formulated as follows:

1. The symbol 1 as a coefficient of a term in the expansion indicates that we must take the entire class represented by the constituent.
2. The coefficient 0 indicates that the corresponding term of the expansion should be discarded.
3. The symbol $\frac{0}{0}$ denotes a completely undetermined portion of the class; that is, we do not know whether to take all, some, or none of its members.
4. All of the other symbols appearing as coefficients of constituents denote that the respective constituents should be set equal to zero; that is, discarded (Reference 327, p. 92).

Let us give an example of how Boole solved a logical problem by using the preceding method. Boole formulated the following problem (Reference 307, pp. 103–104): condition-responsible beings are those rational beings who either possess freedom or have willingly refused it. What can we say about rational beings in terms of responsible beings, possessing freedom or having willingly refused it? Let x denote responsible beings; y, rational beings; z, possessing freedom; and w, having willingly refused it. The conditions of the problem can then be expressed by the equation $x = y \cdot (z + w)$. We must determine y in terms of x, z, w, that is, solve the equation with respect to y.

Boole proceeded as follows. First he solved the equation

$x = y \cdot (z + w)$ with respect to y, using the rules of arithmetic algebra; he obtained

$$y = \frac{x}{z + w}.$$

Then the right-hand side of this expression is expanded with respect to x, z, w according to the formula for expansion of logical functions into constituents, that is, the formula according to which $f(x) = f(1)x + f(0)(1 - x)$:

$$y = \tfrac{1}{2}xzw + \tfrac{1}{1}xz\bar{w} + \tfrac{1}{1}x\bar{z}w + \tfrac{0}{2}\bar{x}zw + \tfrac{1}{0}x\bar{z}\bar{w} + \tfrac{0}{1}\bar{x}z\bar{w} + \tfrac{0}{1}\bar{x}\bar{z}w + \tfrac{0}{0}\bar{x}\bar{z}\bar{w},$$

where the symbol $\bar{x}$ has been introduced as an abbreviation for the expression $1 - x$. Boole then discarded the terms whose coefficients are different from 1 and $\frac{0}{0}$; he set $\frac{0}{0}$ equal to v and thus obtained the equation $y = xz\bar{w} + x\bar{z}w + v \cdot \bar{x}\bar{z}\bar{w}$.

Hence, the solution to the problem says: rational beings are either responsible beings possessing freedom and not having given it up, or those responsible beings who do not possess freedom and have willingly given it up, or, finally, a certain number of irresponsible beings who do not possess freedom and who have not willingly given it up.

Boole did not simply announce his rules of interpretation; he attempted to justify them. Let us consider, for example, the argument by which he justified the fourth of his rules. To this he devoted a special theorem stating that, if any numerical coefficient a of any constituent of an expanded function does not satisfy the fundamental law $a(1 - a) = 0$, the corresponding constituent should be equated to zero (Reference 307, pp. 98–99).*

Boole attempted to prove this theorem as follows. First he supposed that the expansion of the given function (denoted by w) is of the form

$$w = a_1t_1 + a_2t_2 + a_3t_3 + \cdots, \qquad \text{(I)}$$

where $t_1, t_2, t_3, \ldots$ are the constituents and $a_1, a_2, a_3, \ldots$ are the

* According to Boole, the expression x is interpretable in logic only if it satisfies the relationship $x(1 - x) = 0$ ("the criteria of interpretability").

coefficients of the expansion of w. Suppose that $a_1(1 - a_1) \neq 0$; then $a_2(1 - a_2) \neq 0$, while $a_3(1 - a_3) = 0$, $a_4(1 - a_4) = 0$, and so on.

Squaring both sides of Equation I and taking into account the property of the constituents of an expansion according to which $t_i \cdot t_j = 0$ for $i \neq j$ and $t_i^2 = t_i$ (w^2 must equal w), we obtain

$$w = a_1^2 t_1 + a_2^2 t_2 + \cdots \tag{II}$$

Subtracting II from I we have

$$(a_1 - a_1^2)t_1 + (a_2 - a_2^2)t_2 = 0,$$

which is equivalent to

$$a_1(1 - a_1)t_1 + a_2(1 - a_2)t_2 = 0.$$

Multiplying both sides of this equation by t_1 and taking into account the relationship $t_1 \cdot t_2 = 0$, we obtain $a_1(1 - a_1)t_1 = 0$, from which we can conclude that $t_1 = 0$ since we have already assumed that $a_1(1 - a_1) \neq 0$. In exactly the same way, we can conclude from $a_2(1 - a_2) \cdot t_2 = 0$, where $a_2(1 - a_2) \neq 0$, that t_2 is also equal to zero. In Boole's opinion this completes the proof.

Without going into a very deep analysis of Boole's proof, we can show that it is erroneous. In the last step of the proof, Boole had an equation of the form $a_i \cdot (1 - a_i)t_i = 0$, where $a_i \cdot (1 - a_i) \neq 0$ is given as a condition. From this, Boole concluded that $t_i = 0$. It is clear, however, that such a deduction holds only in systems where there are no divisors of zero, that is, where the equation $x \cdot y = 0$ legitimately does imply that at least one of the multipliers must be equal to zero.

It is also clear that in systems with divisors of zero, Boole's proof is simply untrue; in fact, the equation $xy = 0$ can occur in the algebra of logic, with $x \neq 0$ and $y \neq 0$ simultaneously.

For example, the class of "prime numbers" and the class of "numbers divisible by 6" do not intersect (they give the intersection "zero," since prime numbers cannot be divided by 6), yet both classes are obviously nonempty.

In addition, it would seem that this theorem of Boole's is

generally false. In fact, the product xy, where x and y are zero or one, must itself be equal to zero or one; the same is true of the sum $x + y$ since $1 + 1 = 0$. However, every function is obtained by means of successive application of the operations of addition and multiplication. Thus, the value of a function whose arguments can take only the value 0 or 1 should also be only 0 or 1. Those coefficients in Boole's expansion that are different from 0 and 1 simply cannot exist.

How, then, can there appear, in the expansion, coefficients different from 0 and 1, that is, for which the equation $x(1 - x) = 0$ does not hold?

Boole closed his eyes to this circumstance when he carried out division completely formally, without any attempt at all to verify it and, in this case, did not apply the rules he himself had established. We will have more to say about this in Section 5.

Next, Boole introduced a series of supplements to the method of elimination he has presented; this is necessary because the formal device is directly applicable only to logical equations taken individually (and not to systems of logical equations!). In fact, in logic we often have to deal not with one, but with several logical equations which appear as premises. The problem then arises of reducing a system of premises to one premise which can be considered equivalent to that system. In order to solve this problem, Boole proposed a number of methods. One consists of a certain procedure for transformation using a so-called undetermined multiplier. However, this procedure somewhat complicates the process of elimination itself. Thus, it is more convenient to solve a series of equations by using the following rules proposed by Boole:

1. Since any equations of the form $v = 0$ are such that v satisfies the fundamental law $v(1 - v) = 0$, we can combine a set of such equations by simple addition.
2. For equations of the form $v \neq 0$, we can square both sides of the equations, thus reducing them to equations for which Rule 1 holds (Reference 327, p. 123).

An equation of the form $X = vY$ was written by Boole in the equivalent form $X(1 - Y) = 0$. To express in "zero form" the equation $X = Y$ (this can be considered a statement in which both the subject and the predicate are distributed), Boole carried out the following interesting method. First, using the principles of elementary algebra, he transferred Y from the right-hand side of the equation $X = Y$ to its left-hand side, changing the sign of Y from positive to negative: $X - Y = 0$. Naturally, this procedure is logically unjustified. Then he squared both sides of the preceding equation, obtaining $X - 2XY + Y = 0$ or

$$X(1 - Y) + Y(1 - X) = 0,$$

which can be logically justified by interpretation of $\cdot$ as the symbol for conjunction, $+$ as the symbol for strict disjunction, the expression $1 - Y$ as an abbreviation for "not-Y," and the symbol 0 as denoting the identically false constant. Since $(1 - X)$ is equivalent to $(1 + X)$, the equation $X(1 - Y) + Y(1 - X) = 0$ is equivalent to $X(1 + Y) + Y(1 + X) = 0$, or $X + XY + Y + YX = 0$; or, since $XY + YX = 0$, we obtain $X + Y = 0$, which now expresses the idea that the functor of strict disjunction is logically contradictory with respect to the functor of logical equivalence. Thus, the equality $X = Y$ is equivalent to the equality $X + Y = 0$. In other words, Boole had again used an illegitimate method to arrive at a completely justified result.

By using an analogous device, Boole transformed an equation of the form $vX = vY$ into the equation $vX(1 - Y) + vY(1 - X) = 0$.

In considering the technical aspects of the process of elimination, Boole deemed it possible to classify this process as syllogistic by its formal nature. However, in his opinion, deduction cannot in all cases be reduced to processes of elimination.

4. Boole in the Estimation of His Contemporaries

Boole's logical accomplishments were not very well understood by his contemporaries, who could not comprehend the idea of the Boolean operation of addition* and considered Boole's logical

* Boole's definition of addition was directed toward the construction of a logical calculus by analogy with arithmetic. The first such construction was achieved by the Soviet mathematician I. I. Zhegalkin, who showed the existence of a relationship between a Boolean algebra and rings of characteristic two (Reference 315).

system artificial and arbitrary as a whole. This is understandable, for Boole himself was unaware of the connection between the operation of strict disjunction and addition modulo two, although he made *de facto* use of a binary arithmetic. It did not escape the attention of certain of Boole's critics (for example, P. S. Poretskiy) that Boole sometimes simply adjusted logical operations to fit arithmetic ones, not revealing the actual secret of the success of his method and not investigating sufficiently why and to what extent his analogies held.

Boole's pupil Stanley Jevons (1835–1882) characterized his teacher's system as an amazing combination of truths and mistakes (Reference 326, p. 116).

Jevons felt that the Boolean definition of the operation of addition, $x + y$, in proposing the exclusion of the common portion of the classes x and y, entailed insurmountable difficulties since it was unclear, in this case, how to interpret expressions of the type $x + x$.

In his criticism of Boole's understanding of the operation of addition, Poretskiy took a more moderate position than Jevons. This is evident from the following remark by that pioneer of mathematical logic on Russian soil [the remark concerns Boole's operation of addition — N.S.]: "This incorrectness in Boole is of such a nature that one can be reconciled to it without any particular damage occurring" (Reference 316, p. 29).

There exists the opinion that all of the representatives of ordinary logic, without exception, reacted negatively to Boole's results, being of the opinion, with Lotze, that only the premises and the final equation were logical facts and that the intervening procedure was a meaningless paradox (Reference 317, pp. 256–259). However, this point of view does not correspond to reality. For example, the Kiev logician F. Kozlovskiy (Reference 318) gave a positive evaluation of Boole's results. The following statements by Kozlovskiy might be of interest: "It is difficult to agree with Delboeuf's opinion that Boole, by chance, has managed to obtain valid results by false methods: it is not possible that such a complicated system, being in its foundations erroneous, would be valid in all of its conclusions" (Reference 318, p. 36). Later on, the author noted: "Boole's system can be regarded as a particular application of the theory of probability. The theory of probability is concerned with the analysis of those facts whose assertion or negation is not . . . logically necessary. But in a special case, if the probability of the facts is equal to 0 or 1, the theory of probability becomes a theory of certainty; the analysis of certain, that is, logically necessary, facts is also the concern of formal logic" (Reference 318, p. 37).

For his time, the opinion of Kozlovskiy, a student in the department of history and philology at Kiev University, was very bold, and is worthy of our attention, especially in comparison with Lotze's criticism.

We should briefly note the opinion of the well-known propagandist for symbolic logic, John Venn (1834–1923). He stated that most readers who glance through Boole's book do not gain an understanding of the processes of proof used in it (Reference 319, pp. 480–481). Venn compared the relationship between logical multiplication and division in Boole's calculus with the relationship existing between differentiation and integration in classical mathematical analysis (Reference 328, p. 403). However, Venn did not manage in Reference 319 to reveal the secret of success of the method of his great predecessor, and he came to completely pessimistic conclusions.

Can Boole's method be rigorously verified and not simply justified on the basis of the success of its final results?

In considering this question, Venn noted that, at the given level of knowledge about the nature and operation of our mental faculties, evidently a definitive answer cannot be given (Reference 319, p. 485). But later (Reference 328) Venn was actually able to reveal the secret of the success of Boole's procedure, by carrying out a logical analysis of the operation of division (for classes). It was he, in particular, who obtained the relationship analogous to our Formula XVI in Section 5. However, Venn did not consider the problem of carrying out the corresponding corrections in Boole's system. We will discuss this at the end of Section 5. (See also Reference 363.)

5. On the Nature of the Procedure in Boole's Calculus

Let us now consider the problem of elucidation and verification of the "secret" Boolean procedures presented in the preceding sections. In the contemporary literature on the history of logic there is no clear and comprehensive solution to this problem. Boole is usually reproached for introducing into his calculus meaningless symbols, for example, the constants $\frac{1}{0}$ or $\frac{0}{0}$. There exists an opinion that all attempts to justify Boole's methods should come along the line of application to them of the concepts of multivalued logic. In such case, the Boolean symbol $\frac{1}{0}$ could naturally be put in correspondence with the mode "impossibility" and the symbol $\frac{0}{0}$ with the mode "undetermined" (see, for example, References 309 and 322, p. 112).

While we do not quarrel with the principle of such an approach, we think that it is possible to proceed by a simpler route. This method does not require us to leave the framework of a two-valued formalism or to extend the Boolean ring. We note that if the so-called Boolean algebra is to be embedded in the ring of residues modulo 2 of I. I. Zhegalkin (Reference 315) and M. Stone (Reference 323), then the embedding, in the ring, of an algebra of logic in the form presented by Boole himself (using, in particular, the operation of division) is as yet an unsolved problem.

As can be easily shown, the general canonical equation to which any expression in the algebra of logic can be reduced is

$$Ax + B(1 - x) = 0, \tag{I}$$

where $+$ is the symbol for strict disjunction, corresponding to addition modulo 2. By means of equivalent transformations in the ring, Equation I can be expressed in the form*

$$x(A + B) = B. \tag{II}$$

By using one of the Boolean definitions of the inclusion relation —

* See, for example, my article (Reference 324). Equation I has been studied by E. Schröder (Reference 325) where, in particular, he proved that a necessary condition for the solvability of Equation I is the equation $AB = 0$. Additional study of Equation I was carried out in 1895–1896 by the Russian logician and astronomer P. S. Poretskiy (References 321 and 314).

namely, that according to which $a \leqslant b$ if and only if there exists some class v such that $a = v \cdot b$ — we can state that Equation II directly implies

$$B \leqslant (A + B). \tag{III}$$

(Here v plays the role of x in Equation II). Thus, Relationship III is a necessary and sufficient condition for solvability of Equation II. Now we note that Condition III is equivalent to the condition

$$AB = 0. \tag{IV}$$

We prove that III implies IV as follows: multiplying both sides of Expression III by $(1 + A + B)$, we obtain

$$B(1 + A + B) \leqslant (1 + B)(1 + A + B) = 0$$

or

$$(B + AB + B) \leqslant 0, \tag{V}$$

that is, $AB \leqslant 0$. According to the definition of inclusion introduced, from $AB \leqslant 0$ it follows immediately that there exists a v such that $AB = v \cdot 0$, that is, $AB = 0$. Conversely, from IV we can derive III. Adding B to both sides of IV, we obtain

$$B = B(A + 1). \tag{VI}$$

Multiplying both sides of VI by $A + B$ and using the fact that, by assumption, $AB = 0$, we then obtain $(A + B)B = B$, which, by the definition of the inclusion relation, yields III (here B plays the role of v in the relationship $B = B \cdot (A + B)$. In general, the definition of inclusion introduced above directly implies that a necessary and sufficient condition for solvability of the equation

$$Mx = N \tag{VII}$$

is the relation

$$N \leqslant M. \tag{VIII}$$

It is obvious that one of the roots of Equation VII will be the expression N. Actually, if $N \leqslant M$, there exists a v such that $N = v \cdot M$. If, in the formula $MN = N$, we replace N with the expression vM, we obtain $M(v \cdot M) = v \cdot M$, which (by the laws of associativity, commutativity, and idempotence) is equivalent to $vM = vM$, that is, an identity.

Now we note that if x_1 is a root of Equation VII, any expression of the form $x_1 + v(1 - M)$ is also a root, for

$$M(x_1 + v(1 - M)) = Mx_1 + (1 - M) \cdot M = Mx_1 = N.$$

Now we will prove that *every* root x of Equation VII can be represented in the form $x_1 + v(1 - M)$ (for some v).

Thus we have

$$Mx = N \tag{IX}$$

and

$$Mx_1 = N. \tag{X}$$

Adding Equations IX and X termwise, we obtain

$$M(x + x_1) = 0. \tag{XI}$$

Adding $(1 - M) \cdot (x_1 + x)$ to both sides of this equation, we obtain

$$M(x + x_1) + (1 - M) \cdot (x + x_1) = (1 - M)(x + x_1) \tag{XII}$$

or $x + x_1 = (1 - M)(x + x_1)$, that is, for v equal to $x + x_1$,

$$x + x_1 = v(1 - M), \tag{XIII}$$

from which, adding x_1 to both sides, we obtain $x = x_1 + v(1 - M)$, which was to be proved.

Since one root of Equation VII is N, it is clear that every root x of Equation VII can be represented in the form

$$x = N + v(1 - M). \tag{XIV}$$

Since, conversely, the expression $N + v(1 - M)$ is, as we have proved, a root of Equation VII, then we can say that, by assigning various values to the parameter v, we can obtain all the roots of Equation VII and only the roots of this equation (assuming, of course, that at least one root exists, that is, that $N \leqslant M$).

We will now illustrate the validity of this assertion by using a pictorial geometric interpretation with Euler circles (see Figure 24). If $N \leqslant M$, that is, if N lies within M, it is clear that the intersection of M with N is equal to N; that is, $x = N$ satisfies the equation $Mx = N$. If we add to N any class lying in the complement of M, that is, any class represented in the form $v(1 - M)$, then, since the intersection of this latter class with M is empty, we clearly obtain as a result of the addition a class x whose intersection with M is still only N, that is, satisfies the equation $Mx = N$. If we consider some class y, which is not representable in the form

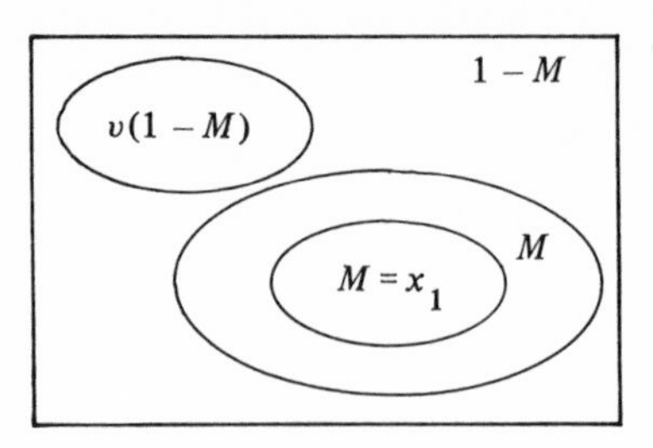

Figure 24

$N + v(1 - M)$, that is, whose part p lying in M (in particular cases, P may equal 0 or y) is different from N, then we have $My = P$ and $P \neq N$; in other words, y is not a root of the equation $Mx = N$. Since Formula XIV represents any root of the equation $Mx = N$ and since Boole always represents such a root in the form

$$x = \frac{N}{M}, \tag{XV}$$

then we must naturally equate (by definition) the expression $\frac{N}{M}$ with the expression $N + v(1 - M)$. We must also assume that, in order to obtain a root of Equation VII, the relationship $N \leqslant M$ must be satisfied and the parameter v must be assigned some fixed value. In order to indicate that $\frac{N}{M}$ depends on v, we will write $\frac{N}{M}(v)$. Thus we have

$$\frac{N}{M}v \mathrel{\overline{\underline{\circ}}} N + v(1 - M). \tag{XVI}$$

(The symbol $\overline{\underline{\circ}}$ expresses the fact that the left side is only another way of denoting the right.)

Equation XVI is the key to the riddle of Boole's procedure. Thus, the operation of division (in the sense of Equation XVI) can be used in logical calculus, precisely because the fractional notation can be eliminated (by use of Formula XVI). The formula for expanding logical expressions into constituents, that is, the formula according to which we have $f(x) = f(1)x + f(0)\cdot(1 - x)$, can now be applied to $\frac{N}{M}(v)$ as well, specifically to the representation of this fraction in the form $N + v(1 - M)$.

Let us return now to consideration of the equation $x(A + B) = B$, which has roots under the condition $B \leqslant (A + B)$. Now we have the right to say that

$$x = \frac{B}{A + B}(v). \tag{XVII}$$

Applying Formula XVI to Expression XVII, that is, setting $M = A + B$; $N = B$), we obtain

$$\frac{B}{A + B}(v) = B + v\cdot(1 + A + B). \tag{XVIII}$$

This is the general expression for the root under the condition that $B \leqslant (A + B)$. Holding v constant and expanding the right-hand side of Equation XVIII with respect to A and B, we obtain

$$\begin{aligned} B + v(1 + A + B) = {} & (1 + v)(AB + O\cdot A\cdot(1 - B) \\ & + 1\cdot(1 - A)B + v(1 - A)(1 - B). \end{aligned} \tag{XIX}$$

By expanding

$$\frac{B}{A + B}(v)$$

formally into constituents, we would obtain

$$\begin{aligned} \frac{B}{A + B}(v) = {} & \frac{0}{0}(v)AB + \frac{0}{1}(v)A(1 - B) + \frac{1}{1}(v) \\ & \times (1 - A)B + \frac{0}{0}(v)\cdot(1 - A)(1 - B). \end{aligned}$$

But, by the definition, we have that

$$\frac{1}{0}(v) = 1 + v(1 - 0) = 1 + v; \quad \frac{0}{1}(v) = 0 + v\cdot(1 - 1) = 0;$$

$$\frac{1}{1}(v) = 1 + v\cdot(1 - 1) = 1; \quad \frac{0}{0}(v) = 0 + v(1 - 0) = v;$$

that is, the result coincides exactly with what Boole needed since the coefficient $\frac{1}{0}(v)$, that is, $1 + v$, stands before the constituent AB, which is equal to zero, on the strength of the assumption that $B \leqslant (A + B)$, a necessary condition for the solvability of Expression XVII. Hence, the term $(1 + v) \cdot AB$ is identically equal to zero. Thus, Boole could use fractional coefficients in his expansions: their interpretation does not present any difficulties.

In connection with the fourth of Boole's rules of interpretation and the "theorem" corresponding to it (see Section 3 of this chapter), we note only that, in essence, Boole had the right to use any natural numbers, as long as he equated them with their residues modulo 2; that is, in his system there should actually be only two numbers, 0 and 1; the fourth of his rules of interpretation can therefore, in fact, relate only to the "fraction" $\frac{1}{0}$. Evidently Boole envisioned, for each expression of the form $\frac{a}{b}$, a corresponding solvable equation $bx = a$, that is, with the "numerator" of the fraction not exceeding its "denominator." But in the fraction $\frac{1}{0}$ this requirement is not met (the constituent with coefficient $\frac{1}{0}$ has, in the "expanded" form, the coefficient $1 + v$, that is, the complement of v). Therefore, Boole usually formulated his problems in such a way that, in their solutions, there will not be constituents with both coefficients: v and also $1 + v$. To avoid such a pair, it is sufficient to add to the problem (if the corresponding condition did not already follow from its formulation) the condition that at least one of the constituents with coefficients v or $1 + v$ be equal to zero. Most of the time Boole did this; sometimes, however, he did it only implicitly. Before presenting an example, we note only that Rule 4 for interpretation of the result simply identifies what is desired with what is attained (hence, the corresponding "theorem.").

As an example we will consider the problem posed by Boole and discussed in Section 3 of this chapter. We saw that Boole eliminated the terms with the coefficients $\frac{1}{2}$, $\frac{0}{2}$, and $\frac{1}{0}$, in accordance with Rule 4 for interpretation. In correspondence to 2-arithmetic, two of these are terms with coefficient $\frac{1}{0}$ (that is, $1 + v$), and one is a term with coefficient $\frac{0}{0}$, (that is, v). The terms with coefficients $\frac{1}{0}$ are discarded on the strength of the condition for solvability of the problem: $x \leqslant (z + w)$ (we leave it to the reader to prove that this condition implies the equalities $xzw = 0$ and $x\bar{z}\bar{w} = 0$). The term $v \cdot \bar{x}zw$ is discarded only on the strength of the supplementary condition that z and w are mutually exclusive (the properties "possess freedom" and "have willingly refused it"), that is, the class zw is empty. Boole did not stipulate this condition; the corresponding premise is actually implied by him, although it is hidden because of the application of Rule 4 (which is why he wrote this coefficient not as $\frac{0}{0}$, but as $\frac{0}{2}$).

The sorts of correction which we have introduced into Boole's method do not prevent us, however, from coming to the conclusion that his method is, in essence, justified; we have revealed only the secret of its success. Of course, until after the work of I. I. Zhegalkin, who actually validated the Boolean operation of addition and embedded the Boolean algebra (which did not exist in the work of Boole himself) in a ring of residues modulo 2 (which was what Boole was actually working with); and, until after the work of J. Venn on Boole, elucidation of the secret of Boole's operation of "division" and the Boolean rules for operating with logical "fractions" was hardly possible.

As far as Boole's actual attempt to construct a logical calculus which would be isomorphic to a field is concerned, it is now clear that this is unrealizable, but only for the following reason: by using logical division we cannot do anything that was not already possible through the use of the operations of addition modulo 2, multiplication, and subtraction from unity; thus, the operation of division can always be eliminated, by use of Formula XVI.

6. Some Further Problems

The analysis of Boole's logical calculus leads to the formulation of many questions whose content would go far beyond the rather limited scope of an evaluation of the ideas and methods of this pioneer in the algebra of logic. Among such questions would be the following: Can one establish rules for operating with logical "fractions" of the form $\frac{x}{y}(v)$* such that a maximal analogy is

* The notation $\frac{x}{y}(v)$ indicates that the expression $\frac{x}{y}$ is a function of the symbol v.

obtained with respect to arithmetic rules for operation with fractions, that is, rules making possible the addition and multiplication of fractions? We are talking about the well-known rules

$$(\Pi_1)\quad \frac{x}{y} + \frac{z}{w} = \frac{xw + yz}{yw} \qquad \text{and} \qquad (\Pi_2)\quad \frac{x}{y}\cdot\frac{z}{w} = \frac{xz}{yw}.$$

It is easy to show that rules (Π_1) and (Π_2) cannot be accepted in the algebra of logic. Let us demonstrate this in the case of (Π_2). The formula

$$(F_2)\quad \frac{x}{y}(v)\cdot\frac{z}{w}(v) = \frac{xz}{yw}(v),$$

where $\cdot$ is the symbol for logical multiplication and v is the symbol for an arbitrary class, cannot be accepted. In this case, we would have

$$\frac{x}{y}(v) = x + v\bar{y};$$

$$\frac{z}{w} = z + v\bar{w};$$

$$\frac{x}{y}(v)\cdot\frac{z}{w}(v) = (x + v\bar{y})(z + v\bar{w}) = xz + v\bar{y}z + vx\bar{w} + v\bar{y}\bar{w}.$$

On the other hand we have,

$$\frac{xz}{yw}(v) = xz + v\bar{y}\bar{w} = zx + v\bar{y} + v\bar{w} + v\bar{y}\bar{w}.$$

The expressions

$$xz + v\bar{y}z + vx\bar{w} + v\bar{y}\bar{w} \tag{1}$$

and

$$xz + v\bar{y} + v\bar{w} + v\bar{y}\bar{w} \tag{2}$$

are not identically equal. We can convert Expression 2 to the form

$$xz + v\bar{y}z + v\bar{y}\bar{z} + vx\bar{w} + v\bar{x}\bar{w} + v\bar{y}\bar{w}. \tag{2'}$$

Comparing this with Expression 1, we see that Expression 2′ has extra terms, namely, the "constituents" $v\bar{y}\bar{z}$ and $v\bar{x}\bar{w}$.

In a similar way, we can show that the relationship

$$(F_1) \quad \frac{x}{y}(v) + \frac{z}{w}(v) = \frac{xw + yz}{yw}(v)$$

is also unacceptable in the algebra of logic.

Nevertheless, we would like to extrapolate Relationships Π_1 and Π_2 in some form (that is, with certain modifications) to the algebra of logic. That is to say, we would like to structure logical calculus (in a ring of residues modulo 2) in a manner maximally analogous to an arithmetic field. Previously, we showed that the introduction of division into a ring of residues modulo 2 is not, so to speak, essential since any relationships using the operation of division can also be obtained through the operations of symmetric difference, multiplication, and subtraction from unity. Thus, as has been demonstrated, division can always be eliminated in a logical calculus. Hence, even if we succeed in introducing into the logical calculus some modification of the relationships (Π_1 and Π_2), the calculus will still not then become a field but will remain a ring.

Now let us formulate (for a Boolean ring) the following problem: "decipher" the symbols $\square$, φ, and λ in the relationships

$$(\mathrm{G}_1) \qquad \frac{x}{y}(v) \cdot \frac{z}{w}(v) = \frac{xz}{yw}(v) \square \varphi$$

and

$$(\mathrm{G}_2) \quad \frac{x}{y}(v) + \frac{z}{w}(v) = \frac{xw + yz}{yw}(v) \square \lambda,$$

where $\square$ is an "empty place" (to be filled by the operations connecting $\frac{xz}{yw}(v)$ with φ and $\frac{xw+yz}{yw}(v)$ with λ), which may be occupied by any symbol for an operation in the algebra of logic, and φ and λ are certain distinct (that is, different from each other) functions in the algebra of logic. Let us begin with (G_1). We found that

$$\frac{x}{y}(v) \, \frac{z}{w}(v) = xz + v\bar{y}z + vx\bar{w} + v\bar{y}\bar{w}. \qquad \text{(XX)}$$

Representing the statement $\overline{T}$ with $1 + T$, we have

$$\frac{x}{y}(v) \cdot \frac{z}{w}(v) = xz + vz(1 + y) + vx(1 + w) + v(1 + y)(1 + w)$$
$$= xz + vz + vzy + vx + vxw + v + yv + wv + ywv. \tag{XXI}$$

On the other hand, we have that

$$\frac{xz}{yw}(v) = xz + v + vyw \tag{XXII}$$

Comparing Equations XXI and XXII, we can write

$$\frac{x}{y}(v) \cdot \frac{z}{w}(v) = \frac{xz}{yw}(v) + v(x + y + z + w + zy + xw). \tag{XXIII}$$

Thus, $\square$ in (G_1) should be replaced by $+$, and φ coincides with $v(x + y + z + w + zy + xw)$. Equation XXIII can be regarded as the analog of the arithmetic rule (Π_2) in the logical calculus constructed as a ring of residues modulo two. Let us now seek the logical analog for rule (Π_1); that is, let us "decipher" the symbols $\square$ and λ in Expression (G_2). We then have

$$\frac{x}{y}(v) + \frac{z}{w}(v) = x + v + vy + z + v + vw \tag{XXIV}$$
$$= x + z + v(y + w).$$

On the other hand, we have that

$$\frac{xw + yz}{yw}(v) = xw + yz + v + vyw. \tag{XXV}$$

Let us put the expression $x + z + v(y + w)$ in the form

$$xw + x\bar{w} + zy + z\bar{y} + vy + vwy + v + v + vw\bar{y}. \tag{XXVI}$$

From a comparison of Equations XXV and XXVI, we conclude that

$$\frac{x}{y}(v) + \frac{z}{w}(v) = \frac{xw + yz}{yw}(v) + x\bar{w} + \bar{y}z + vy + v + vw\bar{y}. \tag{XXVII}$$

Noting that $vy + v + vw\bar{y}$ is equivalent to $v\bar{y}\bar{w}$, we finally obtain

$$\frac{x}{y}(v) + \frac{z}{w}(v) = \frac{xw + yz}{yw}(v) + x\bar{w} + z\bar{y} + v\bar{y}\bar{w}, \qquad \text{(XXVIII)}$$

the desired analog for (Π_1) in the algebra of logic.

In spite of the fact that Relationships XXIII and XXVIII preserve, for operations with logical "fractions," the maximum possible analogy to the rules of ordinary arithmetic, it is preferable to use other, much simpler relationships that can be obtained directly. In fact, given $\frac{x}{y}(v)$ and $\frac{z}{w}(v)$, we can successively obtain

$$\frac{x}{y}(v) + \frac{z}{w}(v) = x + v\bar{y} + z + v\bar{w} = x + z + v(\bar{y} + \bar{w}),$$

from which, by using the tautology $\bar{A} + \bar{B} = A + B$ (for strict disjunction), we have

$$\frac{x}{y}(v) + \frac{z}{w}(v) = x + z + v(y + w). \qquad \text{(XXIX)}$$

Literally: to add two "logical fractions," we must add the sum of their "numerators" to the sum of their "denominators" times some arbitrary indeterminate (v). It is easy to show that Equations XXIX and XXVIII are equivalent. Breaking down the right-hand side of Equation XXVIII, we obtain:

$$\begin{aligned} & xw + yz + v(1 + yw) + x(1 + w) + z(1 + y) + v(1 + y)(1 + w) \\ & \quad = xw + yz + v + vyw + x + xw + z + zy + v + vy + vw + yvw \\ & \quad = x + z + v(y + w), \end{aligned}$$

which is identical to Equation XXIX.

Analogously, given $\frac{x}{y}(v)$ and $\frac{z}{w}(v)$, we can also obtain

$$\begin{aligned} \frac{x}{y}(v) \cdot \frac{z}{w}(v) &= (x + v\bar{y})(z + v\bar{w}) \\ &= xz + zv\bar{y} + xv\bar{w} + v\bar{y}\bar{w} \\ &= xz + v(z\bar{y} + x\bar{w} + \bar{y}\bar{w}). \end{aligned}$$

Thus, we have

$$\frac{x}{y}(v) \cdot \frac{z}{w}(v) = xz + v(z\bar{y} + x\bar{w} + \bar{y}\bar{w}). \qquad \text{(XXX)}$$

Literally: to multiply two logical fractions, we must add the product of their numerators to the product of an indeterminate (v) times the sum of the following three terms: (1) the product of the "numerator" of the first fraction and the negation of the "denominator" of the second; (2) the product of the "numerator" of the second fraction and the negation of the "denominator" of the first; (3) the product of the negations of both "denominators."

It is easy to show that Relationship XXX is equivalent to Relationship XXIII. Indeed, breaking down the right-hand side of XXX, we successively obtain

$$\begin{aligned} xz + v(z\bar{y} + x\bar{w} + \bar{y}\bar{w}) \\ &= xz + vz(1 + y) + vx(1 + w) + v(1 + y)(1 + w) \\ &= xz + vz + vzy + vx + vxw + v + vy + vw + vyw \\ &= xz + v + vyw + v(x + y + z + w + zy + xw) \\ &= \frac{xz}{yw}(v) + v(x + y + z + w + zy + xw), \end{aligned}$$

that is, Equations XXX and XXIII are equivalent.

The problems of the characteristics of Boole's system as a whole are of great interest. Garrett Birkhoff sees in Boole's logical calculus a distributive structure with 0 and 1 and complementation. One should add to this that Boole, in fact, operated with a four-valued model of a distributive structure with complementation. The elements of this model are his coefficients $\frac{1}{1}$, $\frac{0}{1}$, $\frac{1}{0}$, and $\frac{0}{0}$. The third coefficient is the negation of the fourth. The expression $\frac{0}{0}$ can be replaced by v, the fraction $\frac{1}{0}$ by $1 + v$; but v and $1 + v$ are mutually complementary.

Boole's work is a far-reaching generalization of Aristotelian syllogistics; Boole introduces complementation of terms, the universal class, and the null class, and actually operates with a four-valued model of a distributive structure. The special logical problems that arise in the analysis and verification of the Boolean procedures presented in our outline could be supplemented by the formulation of many questions of a methodological nature, for example, the following: To what extent does approximation by a Boolean ring simulate the actual mental reactions realized in using the logical constants "or," "and," "not," and "except"? There are many other interesting types of questions which could be posed as well.

CHAPTER SIX

The Development of the Algebra of Logic After Boole at the End of the Nineteenth Century

1. Systematization and Further Advances in the Algebra of Logic by W. S. Jevons and E. Schröder

The logical results obtained by George Boole underwent further development and generalization in the works of his countryman, the philosopher, logician, and economist William Stanley Jevons (1835–1882). Jevons was educated in the humanities and became interested in mathematics somewhat later, after having completed his schooling. For five years he worked as a chemist at the mint in Australia. From 1866 to 1876, Jevons was professor of Logic, philosophy, and political economy at the University of Manchester; during the period 1876–1880, he was professor of political economy at London. In 1882 he drowned while swimming in the sea.

Jevons' fundamental works in logic are the following: *Principles of Science* (London, 1873), *Elementary Lessons in Logic* (London, 1870), *Pure Logic and Other Minor Works* (London and New York, 1890), *On the Inverse or Inductive Logical Problem* (Manchester, 1876), *Who Discovered the Quantification of the Predicate?* (1873), *On the Mechanical Performance of Logical Inference* (1870), *The Foundations of Logic* (Russian translation by E. Debol'skiy, St. Petersburg, 1878). Jevons' *Principles of Science* also exists in a Russian translation (M. A. Antonovich, St. Petersburg, 1881). For further information on Jevons' theory of logic, see References 371–379.

According to Jevons, the fundamental aim of logical theory consists in finding effective means of solution of so-called logical problems. In this context, Aristotelian syllogistics must be viewed as a theory applicable only to a particular case. In the general problem, we must, on the basis of information formulated in certain conditions, obtain exhaustive information relative to certain logical classes involved in the conditions of the problem. Jevons denoted these logical classes by using small Latin letters $a, b, c, d, \ldots$.

For n letters $a, b, c, d, \ldots$, we can form an initial logical alphabet which consists of 2^n members: the "alternatives" (Jevons' term) represented by all possible classes formed from the given classes $a, b, c, d, \ldots$ or their complements, by the operation of intersection. If we are given a finite number Q of premises, then a certain portion of the alternatives in the initial logical alphabet will be found to be incompatible with these premises. These alternatives can be eliminated by a process of carrying out $2^n \cdot Q$ separate comparisons (by comparing each premise with all of the possible alternatives). After eliminating the alternatives which contradict the premises, we obtain a "simplified logical alphabet," which, like the initial alphabet, is a homogeneous function of n variables. For example, in the case of three simple classes a, b, c, the initial logical alphabet takes the form

$$abc,\ \bar{a}bc,\ \bar{a}\bar{b}c,\ ab\bar{c},\ a\bar{b}c,\ a\bar{b}\bar{c},\ \bar{a}b\bar{c},\ \bar{a}\bar{b}\bar{c}. \tag{1}$$

If we are given the $Q = 2$ premises $a \leqslant b$ and $b \leqslant c$, then Jevons' method requires $2^3 \cdot 2 = 16$ separate comparisons, as a result of which we can eliminate from Form 1 the combinations

$$ab\bar{c},\ a\bar{b}c,\ \bar{a}b\bar{c},\ a\bar{b}\bar{c}.$$

Thus, the simplified logical alphabet is of the form

$$abc,\ \bar{a}bc,\ \bar{a}\bar{b}c,\ \bar{a}\bar{b}\bar{c}. \tag{2}$$

However, it is easy to see that, for large n, Jevons' method becomes very tedious to apply. For $n = 5$, $Q = 4$, we must already carry out 128 separate comparisons.

In Jevons' opinion, any statement can be expressed through an identity relation between its subject and its predicate. Therefore, he used the equality sign to denote the copula in a statement. Jevons omitted inverse operations (subtraction and division) in his logic, limiting himself to direct operations (addition, multiplication, negation). In place of Boolean strict exclusive disjunction, Jevons used inclusive disjunction. Here classes are denoted by capital italic letters; their complements relative to the universal class, by small italic letters. The law of contradiction in the calculus of classes is denoted by the equation $Aa = 0$, where 0 is the symbol for the null class. The rule for expanding logical expressions into components can be written in the form

$$A = A(B + b)(C + c).$$

Jevons distinguished three types of identity relations between classes: (1) simple identity (symbolically: $A = B$), (2) partial identity ($A = AB$), (3) limited identity ($AB = AC$). The following provides an example of limited identity: "Mercury (B) is a liquid metal (C) at low temperature (A)."

Jevons introduced the concept of the *type* of a Boolean function. If one function can be obtained from another by means of permutation of variables and replacement of certain letters by their negations, these two functions are said to be of the same type. For example, $f(x, y, z)$ and $f(\bar{y}, x, z)$ are functions of the same type. The importance of this concept lies in the fact that, as a result, it was discovered that uniform (with respect to type) logical functions can be physically realized by equivalent relay-contact circuits.

Jevons also demonstrated the applicability of the theory of types of Boolean functions to heuristic arguments long before D. Poy.

The theory of induction advanced by Jevons is also of some interest. He states that every scientific induction is nothing more than an application of the inverse method of the mathematical theory of probability. For example, suppose that, as reasons for a given series of events W, we can consider, *a priori*, the reasons $A_1, A_2, A_3, \ldots, A_{n-1}, A_n$ such that

$$A_1 = x_1, \quad A_2 = x_2, \quad A_3 = x_3, \ldots, A_{n-1} = x_{n-1}, \quad A_n = x_n.$$

The probability of event w under the condition A_1 is equal to v_1; the probability of event w under the condition A_2 is equal to v_2;

and, in general, the probability of event w under the condition A_m is equal to v_m. Then we have

$$x_m = \frac{v_m}{v_1 + v_2 + \cdots + v_m + v_{m+1} + \cdots + v_n}.$$

Jevons reduced inductive argumentation to the construction of a certain number of hypotheses. He boldly recommended that this generalization be carried out *a priori*; then the obtained results can be verified by comparison of the hypothetical consequences with the actual data. Suppose, for example, that in n experiments an observed event A corresponds to the presence of another event B. The probability of A and B occurring simultaneously in the first experiment can be set equal to $\frac{1}{2}$; then the probability of coincidence of A and B in n experiments is $(\frac{1}{2})^n$. If A and B are necessarily linked, the probability of their mutual occurrence in any arbitrary number of experiments remains the same, namely, 1. If

$$\lim \left(1 - \frac{1}{1 + (\frac{1}{2})^n}\right) = 0$$

as $n \to \infty$, then A and B are necessarily linked. For $n = 100$, we have

$$\frac{1}{1 + (\frac{1}{2})^{100}}.$$

The difference

$$1 - \frac{1}{1 + (\frac{1}{2})^{100}}$$

is close to zero, from which it follows that A and B are necessarily linked.

The problem of modeling logical operations was also investigated by Jevons. He constructed a "logical machine" whose principle of operation is based on the analogy between the logical process of excluding combinations of classes which are incompatible with the premises and the mechanical action of levers (Reference 380, p. 517). Jevons' machine is preserved in the museum of science at Oxford. It resembles a miniature piano whose keyboard has 21 keys. On sixteen of the keys are written letters, symbolizing the terms in the premises. The letters on the left-hand side of the identity sign in the statements (the so-called subject letters) are marked on the left-hand side of the keyboard. Likewise, the letters on the right-hand side of the identity sign (the so-called predicate letters) are on the right-hand side of the keyboard. The remaining five keys represent operations used to express the copulas, stop

Stop	Left-hand side of statements										Copula	Right-hand side of statements									Stop
	.	.	*d*	*D*	*c*	*C*	*b*	*B*	*a*	*A*		*A*	*a*	*B*	*b*	*C*	*c*	*D*	*d*	. .	
	Or																			Or	

Figure 25

commands, and disjunctive unions that occur in the statements. The entire keyboard looks like that shown in Figure 25.

The machine is operated by pressing down the keys in order, as indicated by the series of letters in the given logical equation. After having thus encoded the content of all the premises, the operator will find on the face of the machine the letter combinations compatible with the given premises. In comparison with the design of the machine proposed by Lully, Jevons' "piano" seems to be a much more successful variant for mechanization of the syllogistic process.

The results obtained by Boole and his most immediate successors were collected and generalized by the German mathematician Ernst Schröder. Schröder received his mathematical education at the Universities of Heidelberg and Königsberg. He was professor of mathematics at the Institutes of Technology at Darmstadt and Karlsruhe. Schröder systematized Boole's logical results and presented a detailed exposition of the logic of relations. His methodology is characterized by certain spontaneous materialistic tendencies.

The following works by Schröder are relevant to our discussion: *Vorlesungen über die Algebra der Logik* (Leipzig, 1890–1905), *Der Operationskreis des Logikkalküls* (Leipzig, 1877), *Abriss der Algebra der*

Logik (Leipzig and Berlin, 1910), *Über Algorithmen und Kalkülen* (1877). This latter work was translated into Russian and published in the collection *Fiziko-matematicheskie nauki v nastoyashchem i proshedshem* (Physico-mathematical sciences in the present and in the past), vol. 7, 1888.

Other works by Schröder are *Lehrbuch der Arithmetik und Algebra*, vol. 1 (1873), an article in *Revue de mètaphysique et de morale*, vol. 8 (1900), and *Über das Zeichnen* (Karlsruhe, 1890). For literature concerning Schröder see References 381–393.

Schröder introduced (Reference 381) the term *Logikkalkül* (logical calculus). Here he also used Boole's "law of development" (expansion into constituents) for the purpose of formalizing deductive logic. With Jevons, he considered only the inclusive form of disjunction of statements.

However, Schröder made substantial changes in the form of Boole's logical system (Reference 381). First of all, he placed at the foundation of the calculus of classes not the equality relation, as did Boole, but the relation of set inclusion, which we will write in the form $a \leqslant b$. In addition, the $+$ sign in Schröder's calculus indicates the union of classes without exclusion of their common portion and not the symmetric difference, as it does in Boole's system.

Schröder perfected Boole's methods for reducing a logical expression to normal form. As the normal form for a logical expression denoting a class, he introduced the polynomial p, equivalent to the given expression e and consisting of a sum of products. The individual multipliers in the products were to be either letters without a negation sign or letters with a negation sign over them. By using this normal form, the German mathematician was able to solve the so-called resolution problem for the calculus of classes, inventing for this purpose the method of "dichotomy."

The logical polynomial p is considered to be expanded with respect to the letter a if each of the terms (addends) in p contains either a or $\bar{a}$ as one of its factors, but not both simultaneously.

The method of dichotomy consists in the reduction of a given logical polynomial to its expanded form with respect to a. This is done by replacing the monomial m with the two-term polynomial $ma + m\bar{a}$. To verify an inclusion relation of the form $e_1 \leqslant e_2$, this inclusion is replaced by the inclusion $p_1 \leqslant p_2$, where the polynomial p_1 is equivalent to the expression e_1, and p_2 is equivalent to e_2. Then the polynomial p_1 is expanded with respect to all of the letters occurring in the polynomial p_2. A term m_1 of the expanded expression (which we will denote by p_1^*) is included in a nonzero term m_2 of p_2 if, in the product of m_1 and m_2, no letter occurs together with its negation (that is, if this product is equal to m_1). The expression e_1 is included in the expression e_2 if each term of the polynomial p_1 is included in some term of the polynomial p_2. In Schröder's work, the problem of reducing logical expressions to normal form is closely linked with the accompanying problem of elimination.

Furthermore, we should note that Schröder was the first to discover the fact that (in contemporary language) not every structure is distributive. In other words, he actually made clear that the relationships

$$(ab + ac) \leqslant (a \cdot (b + c)), \qquad (a + (bc)) \leqslant ((a + b) \cdot (a + c))$$

are true for every structure, whereas the converse relationships

$$(a \cdot (b + c)) \leqslant (ab + ac), \qquad ((a + b) \cdot (a + c)) \leqslant (a + (bc)).$$

are not necessarily true for every structure.

Finally, Schröder was the real author of the principle of duality for logical expressions (in the logic of classes). In many works, he considered the problem of elimination in the calculus of classes, for which he proposed the following system of axioms (the symbol $\subset$ indicates set inclusion, and the symbol $+$ indicates union of sets):

1. $x \subset x$;
2. if $x \subset y$ and $y \subset z$, then $x \subset z$;
3. $(x + y) \subset z$ if and only if $x \subset z$ and $y \subset z$;
4. $x \subset (y \cdot z)$ if and only if $x \subset z$ and $x \subset y$;
5. $x(y + z) = x \cdot y + x \cdot z$, where $\cdot$ indicates set intersection;
6. $x \subset 1$, where 1 denotes the universal set;
7. $0 \subset x$, where 0 denotes the empty set;
8. $1 \subset x + x'$, where x' is the complement of x relative to 1;
9. $x \cdot x' \subset 0$.

Later, this axiomatization was simplified to a considerable extent by the American mathematician E. V. Huntington.

Having solved the resolution problem for the calculus of classes, Schröder, refusing the gift of Boole's problematics, developed his own very broad analogy between the devices of classical algebra and the methods of formal logic. In addition to the problem of extracting the consequents and antecedents from logical equations (in the calculus of classes), Schröder also posed the problem of finding the roots of these equations. According to his definition, a root of a given logical equation of the form $Ax + B\bar{x} = 0$ is that value of the letter x whose substitution in the equation $Ax + B\bar{x} = 0$ gives the same expression as would be obtained by eliminating x from the equation (that is, results in the equation $AB = 0$).*

In Schröder's method, as in Boole's, we can derive from

$$Ax + B\bar{x} = 0, \tag{1}$$

the equation

$$x = \bar{A}B + v \cdot \bar{A} \cdot \bar{B}. \tag{2}$$

Schröder obtained this latter equation from the preceding one as follows. First he proved that $Ax + B\bar{x} = 0$ is equivalent to the expression†

$$x = \bar{A}x + B\bar{x}. \tag{3}$$

Now, expanding the right-hand side of Equation 3 with respect to A and B, we obtain

$$x = \bar{A}Bx + \bar{A}\bar{B}x + AB\bar{x} + \bar{A}B\bar{x}, \tag{4}$$

* In the expression $Ax + B\bar{x} = 0$, the + sign is understood in the sense of non-Boolean union (that is, union *including* the common portion of the corresponding classes); the absence of a symbol in the expressions Ax and $B\bar{x}$ indicates set intersection; $\bar{x}$ is the complement of x; and 0 denotes the null class.

† In fact, from $Ax + B\bar{x} = 0$ it follows that $Ax = 0$, which is equivalent to $x \leqslant \bar{A}$ or $x = \bar{A}x$. Since $B\bar{x} = 0$, we then have $x = \bar{A}x + B\bar{x}$. Conversely, from

$$x = \bar{A}x + B\bar{x} \tag{a}$$

we can derive $Ax + B\bar{x} = 0$. For if we multiply both sides of (a) by $\bar{x}$, we obtain $B\bar{x} = 0$. On the other hand, if we multiply both sides of (a) by x, we obtain $Ax = 0$ (taking into account that $AB = 0$). Adding these, we have $Ax + B\bar{x} = 0$.

but, since $AB\bar{x} = 0 \cdot \bar{x} = 0$, we have

$$x = \bar{A}Bx + \bar{A}\bar{B}x + \bar{A}B\bar{x}, \tag{5}$$

which is equivalent to

$$AB \leqslant x \leqslant \bar{A}B + \bar{A}\bar{B}, \tag{6}$$

where the symbol $\leqslant$ denotes the inclusion relation.

Since $x \leqslant \bar{A}B + \bar{A}\bar{B}$ is equivalent to $x = v(\bar{A}B + \bar{A}\bar{B})$, where v is indeterminate,* then we can conclude that Equation 6 is equivalent to

$$x = v(\bar{A}B + \bar{A}\bar{B}) + \bar{A}B. \tag{7}$$

Removing the parentheses on the right-hand side of Equation 7, we find that the last term already contains the first. Thus, we finally obtain

$$x = \bar{A}B + v \cdot \bar{A}\bar{B}, \tag{8}$$

which is the same as the result obtained by solving the equation $Ax + B(1 - x) = 0$ by Boole's method. Let us show that Equation 8 is actually a root of the equation $Ax + B\bar{x} = 0$. Substituting the value of x from Equation 8 into Equation 1, we obtain

$$A(\bar{A}B + v \cdot \bar{A} \cdot \bar{B}) + B(\bar{B}\bar{v} + A) = 0,$$

which yields $AB = 0$.

It now becomes obvious that from a formal point of view there is a difference between a logical root and the root of an algebraic equation. An algebraic root satisfies the corresponding equation identically, whereas a root in logic, when substituted into the original equation, transforms it not into an identity but into the resultant (that is, the result of excluding the letter x from the equation).

Since Equation 8 contains an undetermined class v, Schröder considered the number of roots of a logical equation to be infinite. The Russian successor of Schröder, P. S. Poretskiy, held a different

* We note that in finding roots Schröder used a representation for the inclusion relation which is in agreement with the Boolean definition: $x \leqslant y$ is equivalent to $x = vy$, where v is an undetermined class.

opinion on this matter. In Reference 403, he wrote as follows:* "In determining roots . . ., should we seek for them expressions depending only on the remaining n letters being considered, that is, the letters $b, c, d, \ldots$, or may we resort, for this purpose, to any other letters α, δ, γ which are foreign to the given problem? Schröder . . . was of the second opinion . . . . As for me, I consider that the limits and terms of speech of the problem are binding on us and that, consequently, the roots . . . must not depend on any other letters but the remaining letters in consideration $b, c, d, \ldots$. If we maintain this point of view, then we have to deal with an arbitrary class . . . which is a function of only the n letters $b, c, d, \ldots$. It is known that from n letters we can form only 2^{2^n} different classes. Consequently, the number of roots . . . cannot exceed this number" (Reference 403, p. 589).

We should also note the substantial contribution made by Schröder to the development of a general logical theory of relations. Here Schröder made his discoveries independently of (and often at about the same time as) his American colleague C. S. Peirce (see Section 3 of this chapter). For example, he defined what is meant by a binary relation on an arbitrary class K of elements $x, y, z, \ldots$. As such a binary relation, he had in mind any rule R which would satisfy the condition that for every ordered pair (x, y) of elements in K either xRy holds or $xR'y$ holds, where R' denotes that R is not fulfilled.

Suppose we are given the n elements $x_1, x_2, x_3, \ldots, x_{n-1}, x_n$ of K. Then, according to Schröder, we can set up a one-to-one correspondence between relations R on K and "matrices of relations" of the form $\|r_{ij}\|$, consisting of zeros and ones. This one-to-one correspondence can be defined by use of the conditions

$$r_{ij} = \begin{cases} 1 & \text{for } a_i r a_j, \\ 0 & \text{for } a_i r' a_j, \end{cases}$$

where r' denotes that the relation r is not fulfilled and where

* In Reference 403 the equation $Ax + B\bar{x} = 0$ is written in the form $Ga + Ha_0 = 0$, where a_0 is the complement of a. We seek an expression for a in terms of G and H; hence we impose on the indeterminate v the condition that v cannot depend on letters on which G and H do not already depend.

$\|r_{ij}\| = R_{ij}$, $x_i = a_i$, $x_j = a_j$. It is easy to see that the equality relation E corresponds to the unit matrix $E_{ii} = 1$. Obviously, we have $e_{ij} = 0$ for $i \neq j$. Schröder also defined the following concepts: "null-relation," "composition of relations," "universal relation," and so on. Schröder's investigations in the theory of relations also had a definite influence on the work done by Tarski and McKinsey on the algebra of relations (see, for example, J. C. McKinsey, *Journal of Symbolic Logic*, vol. 5 (1940), pp. 85–97, and A. Tarski, *ibid.*, vol. 6 (1941), pp. 73–89).

According to Alonzo Church (Reference 382), Schröder, in Reference 383, anticipated to a significant degree Bertrand Russell's simple theory of logical types.

The analysis of the foundations of traditional logic (from the point of view of the algebra of logic) also had a definite place in Schröder's work. He very cleverly noted that the differences between certain of the modes of syllogistic figures were actually nonessential, from the point of view of the representation of these modes in the algebra of logic. We will illustrate his idea by comparing the representations of the modes *Darii* and *Datisi* in the specific terminology of the calculus of predicates. The first representation is as follows:

$$\begin{array}{ll} & \exists x(A(x) \cdot B(x)) \\ & \underline{\forall x(B(x) \rightarrow C(x))} \\ \text{Consequently,} & \exists x(A(x) \cdot C(x)). \end{array}$$

The second representation is

$$\begin{array}{ll} & \exists x(B(x) \cdot A(x)) \\ & \underline{\forall x(B(x) \rightarrow C(x))} \\ \text{Consequently,} & \exists x(A(x) \cdot C(x)). \end{array}$$

Actually, the difference between these two representations is no greater than the difference between the algebraic equations $ax = b$ and $xa = b$.

Like W. S. Jevons and W. Clifford, Schröder also considered the problem of types of Boolean functions. He proposed a method for geometric representation of Boolean functions in the form of

n-dimensional hypercubes. A uniform transformation (that is, one preserving type) could then be viewed as the mapping of an n-dimensional hypercube into itself (see the supplement to G. N. Povarov's article "On the group invariance of Boolean functions," in *Primenenie logiki v nauke i tekhnike* [The application of logic in science and technology], 1960, p. 556).

There are some remarkable statements in Schröder's work which are relevant to the beginnings of semiotics. Thus, in Reference 392, in attempting to guarantee the uniqueness of names, Schröder advanced the "axiom of the inherentness of symbols," which postulated the invariability of symbols within the limits of a given formal system. This axiom, wrote Schröder, "gives us assurance that during all our discussions and proofs the symbols will remain fixed in our memory, but even more solidly on the paper" (Reference 392, pp. 16–17).

In addition to Jevons and Schröder, there were other followers of Boole who made important contributions to the development of the algebra of logic. Among them, we should mention Hugh McColl, O. Mitchell, and John Venn.

The first appearance of a pure calculus of propositions dates from 1877 and the publication of articles on logic by Hugh McColl (1837–1909) (see References 394–396). McColl also showed a tendency toward augmenting the two-valued logical formalism with a third truth value; this evoked many objections from P. S. Poretskiy.

According to Peirce, O. Mitchell was the first to introduce the universal and existential quantifiers. In Mitchell's work, the expression $\prod (f(x) = 0)$ denotes that the logical function $f(x)$ is equal to zero for all values of x; the expression $\sum_x (f(x) = 0)$ indicates that the function $f(x)$ is equal to zero for some values of x.

An instructor in logic and morals at Cambridge, John Venn (1834–1923), first used the term "symbolic logic." He revealed the essence of the secret of success of Boole's procedures, thus making an important contribution to the study of the history of symbolic logic (see his book *Symbolic Logic*, London, 1881).

In 1876 Venn put forward the idea of extending the "frequency of occurrence" concept of probability to logic; this was before the appearance of Mises' theory, which also dealt with this idea. Venn's can be considered a direct predecessor of the probability logic of Hans Reichenbach, with whom he shares the "frequency" treatment of the concept of probability. According to Venn, probability logic is the logic of sequences of statements. A single element sequence of this type attributes to the given proposition one of two values: zero or one; an infinite sequence attributes any real numbers which lie in the interval from zero to one.

However, in the development of such a method, certain serious difficulties arise. In an experiment, we always limit ourselves to a finite sequence of statements, with only finite frequency of occurrence, from which we draw a conclusion about the probability of the entire infinite sequence of statements. But this conclusion is itself a hypothesis, whose reliability is a function of the length of the chosen trial segment. Thus, this hypothesis itself, in turn, requires an estimate, and this estimate will then be a hypothesis of second order, and so on. No matter how long a system of hypotheses we construct, the estimate of the last element of such a system will always remain unknown. In addition, we must solve the problem of the possibility of writing out scientific data in terms of a sequence of statements, and this problem is neither clear nor simple. These difficulties, which Venn's theory was not in a position to overcome, explain why Venn's probability logic did not enjoy very wide circulation and why his treatise *The Principles of Empirical or Inductive Logic* (London, 1889) did not become a reference book for those logicians working on inductive problems. Some of the traditional logicians (for example, W. Minto) were shocked by Venn's inclusion of induction in the very definition of the concept of probability.

Venn also attempted to clarify the treatment of a number of questions in traditional logic. In revising the traditional ideas about modality, Venn noted that the exact quantitative evaluation of human certainty is one of the aims of logic, related to the measurement of the modality of a proposition. The question of modality, in Venn's opinion, was related to inductive logic rather than to formal logic since the former considered conditions of various degrees of verisimilitude.

2. The Legacy of P. S. Poretskiy and Ye. L. Bunitskiy

The Russian logician and mathematician, Platon Sergeyevich Poretskiy (1846–1907), professor at the University of Kazan, played an important role in the development of the algebra of logic at the turn of the century. Poretskiy substantially generalized and extended the achievements of Boole, Jevons, and Schröder. The well-known French scientist, L. Couturat, whose book *The Algebra of Logic* was published in 1904, considered Poretskiy's methods to be the culmination point in the development of the algebra of logic during that period. Not only does Poretskiy's work stand at the same level as that of his contemporaries, his work on the algebra of logic can be compared with the corresponding sections of Whitehead and Russell's *Principia Mathematica.*

Even today, Poretskiy's studies continue to exert a stimulating influence on the development of algebraic theories of logic. For example, this influence can be strongly detected in the Ph.D. dissertation of A. Blake, Reference 397. Blake's basic concepts — "syllogistic polynomial" and "simplified canonical form" — are, as the author so eloquently admitted, attempts at combining the method of Boole and E. Schröder with the method of P. S. Poretskiy (Reference 397, p. 2).

Blake considered Poretskiy's method far superior to those of his predecessors. For example, he stated that Schröder's method is unsatisfactory in the sense that it does not completely characterize the class of all inferences which can be drawn from a given logical equation. On the other hand, said Blake, Poretskiy's table of consequents is intended for precisely this purpose (Reference 397, p. 38).

The strong influence of Poretskiy's logical heritage on Blake's investigations was also recognized by Blake's countryman, the well-known logician and mathematician, J. C. McKinsey. In reviewing Blake's dissertation, McKinsey correctly noted that Blake's methods are stimulated by "Poretskiy's tables for finding all the consequents of a Boolean equation" (Reference 398, p. 93).

Poretskiy's work was praised very highly by his contemporaries. For example, in a paper by I. I. Yagodinskiy, we read: "Among the

Russian scientists, Poretskiy worked on mathematical logic with great success" (Reference 399, p. 14).

Unfortunately, the attempt to publish a systematic presentation of Poretskiy's theory of logic in Russia was unsuccessful, primarily because the writer who took this task upon himself, M. Volkov (Reference 426), was not sufficiently qualified. More successful in this respect was L. Couturat. His *Algebra of Logic* was a popular elucidation of some of Poretskiy's methods. Nevertheless, even Couturat's study does not reflect certain very important questions treated in Poretskiy's work, for example, his axiomatization, his theory of canonical forms, his theory of combination and decomposition of logical equations into elements, and so on. Furthermore, the very nature of Couturat's work and the goals he set for himself did not make possible an adequate exposition of the very original content of Poretskiy's conception. It is interesting, for example, that Couturat overlooked Poretskiy's original canonical form, whose application was an essential supplement to the classical method of obtaining the canonical form of an expression (in the algebra of logic).

The focus of Poretskiy's logical investigations was his theory of logical consequents and his original treatment of canonical forms for logical expressions.

Platon Sergeyevich Poretskiy was born October 3, 1846 into the family of an army doctor in Elizavetgrad, the capital of Kherson district. He received his secondary school education in Poltava, after which he studied in the department of physics and mathematics at the University of Kharkov until 1870. At the suggestion of Professor I. I. Fedorenko, he was granted a university stipend in the department of astronomy. In 1873, Poretskiy successfully passed his master's examination. Three years later, he was appointed chief astronomical observer at Kazan University, which was at that time undertaking observation of stars in Kazan's zone for the International Astronomical Society. This rather responsible and, at the same time, laborious task, previously carried out by Professor Koval'skiy, now lay entirely on the shoulders of Poretskiy. Over a period of three years (1876–1879), he took a large number of observations on the meridianal circle. The results of these observations were published in two volumes.

In spite of his poor health, Poretskiy found the time to participate in many societies. After the Physics and Mathematics Section of the Society of Natural Scientists was established at Kazan, he participated very energetically in its work (he was secretary, treasurer, and later a life-time member). For several years he was also editor of the liberal Kazan newspaper *Telegraph*, in which there appeared, at times, his humorous poems and his translations of Béranger.

For his paper "On the Question of Solution of Normal Systems Encountered in Practical Astronomy, with Application to Determination of the Errors in the Divisions of the Meridianal Circle at Kazan Observatory," he was awarded the scientific degree of doctor of astronomy in 1886 and was made assistant professor in spherical trigonometry. In this paper, he undertook a deep theoretical analysis relating to the problem of reducing the number of equations and unknowns in systems of so-called cyclical equations. These are encountered in determining errors in scaling. Poretskiy also carried out the necessary calculations for the numerous observations taken for determining the errors of graduation in the Kazan meridianal circle (Reference 400).

Poretskiy died on August 9, 1907 in the village of Zhoved' in Chernigovskiy District, where he had gone from Kazan after becoming seriously ill. He maintained his lively interest in science and his creative work in the algebra of logic until the last days of his life. He participated, *in absentia*, in many international scientific congresses and carried on an energetic correspondence with Russian, as well as foreign, scientists. At the time of his death, he was working on an uncompleted article in logic.

In 1884, Poretskiy published his major work, "On the Methods of Solution of Logical Equations and on the Inverse Method in Mathematical Logic" (Reference 401), in which he expounded his original theory of logical equations. "Turning to our article," says Poretskiy, "... we should say that: (1) it represents the first attempt (not only in our own but also in foreign literature) at constructing a complete and finished theory of qualitative argumentation, and (2) it represents ... a completely original work, which is of even greater significance because the general formulas and devices of this theory were first obtained by us. An entire portion of this theory (the transition from syllogisms to premises) belongs completely to us, as relates to its devices as well as to the very idea that the solution of such a problem is possible" (Reference 401, pp. XXIII–XXIV).

On the basis of this work, Poretskiy, at Kazan University, became the first person in Russia ever to deliver lectures on mathematical logic.

Poretskiy explained the aim of his article "Solution of the General Problem of the Theory of Probability by Using Mathematical Logic" (Kazan, 1887) as follows: "The solution of this problem given by Boole in his work *An Investigation of the Laws of Thought* cannot be regarded as scientific since it is based on an arbitrary and purely empirical theory of logical equations and also because the very idea of transition from logical equations to algebraic ones was not successfully developed by Boole . . .. The principal aim of the present article is to give scientific form to Boole's deep, but vague and unproved, idea of applying mathematical logic to the theory of probability" (Reference 402, pp. 83–84).

In the article "The Law of Roots in Logic" (Reference 403), Poretskiy solved the elimination problem for logical equations.

In a long article "Sept lois fondamentales de la théorie des égalités logiques" (Reference 404) (1898–1899), he presented a theory of logical equations in systematic and completely finished form. Here Poretskiy formulated the law of consequents and the law of hypotheses relative to the logical justifications for given consequents, laws discovered by him. He also constructed an original and very clever table which allows one to find the consequents as well as the "reasons" (logical justifications) for given logical equations without actually having to resort to complete calculation. However, this device is suitable only for the simplest logical equations; the solution of these problems for systems of logical equations is realized by introduction of his original simplified canonical form (Reference 404).

The article "Teoriya logicheskikh ravenstv s tremya chlenami" [The Theory of Logical Equations with Three Terms], (1901) is a generalization of the classical theory of the syllogism and is of exceptional interest from the point of view of formal logic. Another of Poretskiy's articles, "Iz oblasti matematicheskoy logiki" [From the Realm of Mathematical Logic], (*Fiziko-matematicheskoy ezhegodnik*, 1902, no. 2) is analogous in its content.

In the article "Quelques lois ultérieures de la théorie des égalités logiques" (Reference 405), Poretskiy defined more exactly the theory of hypotheses relative to logical justifications for given consequents. He introduced the concept of a complex logical justification (a "reason," in Poretskiy's terminology).

In the work "Appendice sur mon nouvel travail: Théorie des non-égalités logiques" (Reference 406), Poretskiy briefly summarized and interpreted certain theses in his theory of logical inequalities.

"Théorie conjointe des égalités et des non-égalités logiques" (Reference 407) is devoted to questions concerning the relationships between the calculus of classes and the propositional logic.

Poretskiy was also the author of many other articles and reviews, of which we mention only the following two: "An Elementary Presentation of the Theory of Logical Equations with Two Terms" (Reference 408) and "Novaya nauka i akademik Imshenetskiy" (Reference 404).

In the review "Concerning M. S. Volkov's Handbook 'Logical Calculus'" (Kazan, 1888), Poretskiy upheld his founding role in the development of the theory of logical equations. Here he again emphasized that the central problem in his theory of logic is that of deriving the consequents from a given set of premises and, conversely, of finding those premises from which a given logical equation has been derived as a consequent.

It would be wrong, however, to believe that Poretskiy went beyond the limits of the framework established by Boole and Schröder. He worked entirely within this framework, generalizing its results and "greatly furthering the simplification of the devices originated by Boole and Schröder through his original formulation of the problem" (Reference 410, p. 385). To these words of P. Erenfest one must add, however, that Poretskiy was the first to obtain many results that played a most important role in the emergence of the present-day form of the algebra of logic and are, thus, relevant even today. Also, by using Poretskiy's algorithms it is very easy to solve the Boolean–Schröderian problems of elimination and solution of logical equations.

Let us briefly characterize Poretskiy's theoretical views with regard to cognition. At the beginning of the foreword to his book *On the Methods of Solution of Logical Equations* (Reference 316), the Russian scientist distinguishes between "quantitative forms," studied by algebra, and "qualitative forms," studied by logic. Although this distinction prevents direct application of the methods of algebra to the subject of logic, the well-known correspondence (up to a certain point) between the devices of these two sciences is considered by Poretskiy. This identification makes possible the corresponding modification in algebraic devices and their application to the study of logic. If traditional logic can be linked with verbal expressions which give what might be called a logical "visualization," then mathematical logic, not linked with verbal expression and applying the mathematical method, admits the possibility of greater generalization and of the formulation of more complicated and more specialized logical relationships.

According to Poretskiy, logic should analyze the nature of scientific argument. In such an analysis, the rules of logic would not be independent of the properties of the domain of objects being investigated by the particular science. The law of logic is "a truth, which includes within itself a certain definite indication as to the very nature of the material being investigated" (Reference 401, p. XII).

Therefore, even the algebraic reworking of logic cannot eliminate the question of content. Being a materialist, Poretskiy asserted that any formal logical system constructed axiomatically has the right to exist only if all the statements proved in it are contentually true when applied to some specific domain or aspect of objective reality.

It is, indeed, no accident that Poretskiy dedicated the initial pages of his fundamental work *On the Methods of Solution of Logical Equations* to an interpretation of the substantive content of several of the axioms of his logical system. The professor of astronomy correctly proposed that the formal analytical apparatus of logical calculus is useful only in those cases where it corresponds to a definite real content. In addition, even in the presence of an already constructed axiomatization, substantive understanding of content does not lose its relevance. This, in particular, is very evident in Poretskiy's work; having constructed his original algorithm for elimination, he was not satisfied with the formal side of the question but he went on to strengthen the significance of his device with profound arguments concerning the contentual aspect of the algorithm.

These considerations in no way diminish the principal merits of the algorithm. The point is that construction of an algorithm hardly necessitates its being applied in every concrete case. Since every algorithm concerns not one but a whole class of homogeneous problems, we might foresee the existence of a subclass of complex problems which also could be effectively solved through it. However, it may also be the case that, because of given specific conditions, this subclass of problems can also be solved by a simpler method than application of the algorithm.*

In discussing the antimetaphysical tendencies in Poretskiy's gnosiological views, we will find of interest his report "A Historical Outline of Spherical Trigonometry," pp. 152–167. In this report, the theoretical and cognitive problem of the relationship between the abstract and the concrete in scientific investigation is refracted through the prism of the special question of the effectiveness of the rational formalization of a particular theory. Poretskiy noted that the purely geometric interpretation of spherical trigonometry held by the ancient Greeks (Ptolemy and others) made more difficult its practical use, whereas the more abstract, algebraic form of this science developed by the Arabs of the Middle Ages made possible far more extensive and important practical applications.†

* In concluding our characterization of the materialist foundations of Poretskiy's theory of logic, we should note that, in studying the formal logical rules for deduction, not only did Poretskiy not separate form from content, but he always considered form derivative from content.

† See P. S. Poretskiy, "Istoricheskiy ocherk razvitiya sfericheskoy trigonometrii" [A Historical Outline of the Development of Spherical Trigonometry], *Izvestiya fiziko-matematicheskoy obshchestva pri Kazanskom universitete*, vol. 5 (1887), p. 158. As is now well known, these scientists came mainly from the nations of Central Asia and the Trans-Caucuses: the Khorezmians, the Tadjiks, the Azerbaijanians, and so on.

From these remarks, and also from the general context of the rest of the article, it is clear that Poretskiy was close to understanding the dialectic proposition that the more abstract characteristics of a theory (provided the abstractions employed are really scientific) not only do not weaken its practical implications but actually strengthen them. Thus, the theoretico-cognitive foundations of Poretskiy's logic fall within the general framework of a materialist world view with elements of an antimetaphysical relationship toward the analysis of the processes of development of scientific theories.

The construction of Poretskiy's logical calculus begins with a set of variable elementary terms $a, b, c, d, \ldots$. In addition, there are two constant terms, 1 and 0. These latter terms, interpreted in the objective realm of the calculus of classes, denote, respectively, the universal set ("the world of speech" in Poretskiy's terminology) and the empty set.

Complex terms are formed from elementary ones by use of a sequence of operations: (1) binary operations, denoted by the symbols $\cdot$ and $+$ (in propositional logic these symbols correspond to conjunction and disjunction); (2) unitary operations, corresponding to the formation of the complement (negation — in the propositional logic), for which various notations are used in the course of Poretskiy's work. We shall denote complementation by using a prime following the letter. In the subsequent discussion we shall also use the terms addition and multiplication instead of disjunction and conjunction, respectively.

The fundamental undefined relation, by means of which all the other relations are defined, is the relation of equality. Poretskiy's algorithm is, in fact, a calculus of logical equations. A logical equation is differentiated from any other complex logical class in that it expresses a statement and, furthermore, a necessarily universal one (positive or negative). The equality of particular logical classes means that they are equal in volume (extent).

Then we introduce into consideration the set of all logical equations r of the form $e_1 = e_2$, where $e_1 = \Phi(a, b, c, d, \ldots)$ and $e_2 = \varphi(a, b, c, d, \ldots)$ are logical terms formed by applying the operations $+$, $\cdot$, and $'$ to the terms $a, b, c, d, \ldots$.

By using the equality relation, we can define the relation of inclusion $a \leqslant b_{Df}: \equiv : a = ab$, where the compound symbol $_{Df}: \equiv :$ denotes the equivalence of statements by definition. The letters e_1 and e_2 used here were not introduced by Poretskiy in his calculus; we have introduced them to denote terms and, thus, make commenting on such terms easier. Also, in the future, we shall avoid unnecessary use of brackets; most of the time, we shall do this by applying the ordinary rules of Boolean algebra, although occasionally we shall replace brackets by dots. We shall also omit the multiplication sign when it occurs between letters.

The question of Poretskiy's axiomatization is rather complicated, for Poretskiy nowhere gave a precise enumeration of his axioms. However, there can be no doubt as to the axiomatic character of his logical works. Thus, for example, the Kazan mathematician, N. N. Parfent'ev, asserted that "Poretskiy's calculus is founded on several postulates and then is developed further by means of the deductive–axiomatic route" (Reference 411).

Another Russian mathematician, I. V. Sleshinskiy, correctly assuming that the algebra of logic presented by Couturat (Reference 412) is actually Poretskiy's (in a rather popular and fragmentary elucidation, of course), considered that Poretskiy's axiomatization, in fact, takes the form*

1S. $a < a$,
2S. $((a < b) \cdot (b < c)) < (a < c)$,
3S. $ab < a$,
4S. $ab < b$,
5S. $((a < b) \cdot (a < c)) < (a < b \cdot c)$,
6S. $((a < b) \cdot (b < a)) < (a = b)$,
7S. $(a = b) < (a < b)$,
8S. $(a = b) < (b < a)$.

We assert that the system of axioms and rules 1S–8S is exactly equivalent to the system of axioms for a so-called logical group

* The list of axioms (1S) through (8S) exactly coincides with their listing by I. V. Sleshinsky (see Reference 413).

(Curry's term) since it deals with only one operation (denoted by using the symbol $\cdot$). The system of axioms and rules describing a logical group in terms of the relation of inclusion is as follows (Reference 415, part IV):

1C. $e \leqslant e$,
2C. $((e \leqslant e_1) \ \& \ (e_1 \leqslant e_2)) \rightarrow (e \leqslant e_2)$,
3C. $ee_1 \leqslant e$,
4C. $ee_1 \leqslant e_1$,
5C. $((e \leqslant e_1) \ \& \ (e \leqslant e_2)) \rightarrow (e \leqslant e \cdot e_2)$,
6C. $(e = e_1)_{Df} \colon \equiv \colon (e \leqslant e_1) \ \& \ (e_1 \leqslant e)$.

Axioms 1S–5S coincide exactly with Curry's axioms 1C–5C for a logical group. Axioms 6S–8S are equivalent to Curry's single axiom 6C. We should note that the axiomatization 1S–8S does not draw any distinction between relations and rules. In Axiom 2S for example, the first, second, and fourth $<$ symbols are used as signs for relations, whereas the third $<$ symbol is clearly a symbol for implication (in the sense of a rule of deduction). Similarly, the dot between the first two parentheses in this axiom is a symbol for conjunction in the sense of the propositional calculus and not in the sense of Poretskiy's operation (intersection of classes).

In the propositional calculus, where the difference between operations and relations disappears because the results of operations can also be considered "true" or "false," all of Sleshinskiy's symbols can be interpreted simply. However, in this case it would seem that Sleshinskiy did not formulate the rules of inference from axioms (in Curry's axiomatization Axioms 2C, 5C, and 6C are rules of inference).

Thus, the axiomatization (1S–8S) proposed by Sleshinskiy does not elucidate Poretskiy's actual axiomatization. In his first important investigation, Poretskiy had already introduced axioms and rules corresponding to 1A, 5*A, 6A, 6*A, 7A, 7*A, 9π, $9^*\pi$, 10A, 10*A, 11A, 11*A, 12A, 12*A of the system to be presented here. The remaining axioms in this system can be found in other works by Poretskiy, for example in Reference 404.

We propose as the basis of analysis of Poretskiy's calculus the system of axioms and rules of inference of a distributive structure with complementation, which can be formulated in the following way:

The letters $a, b, c, d, \ldots$ denote variables; the symbols 1 and 0, constants. The operations will be "multiplication," denoted by $\cdot$, addition, $+$, complementation, $'$. "Terms" will be defined, as usual, by induction: (1) the variables $a, b, c, d, \ldots$ and the constants 1 and 0 are terms; (2) if e is a term, e' is also a term; (3) if e_1 and e_2 are terms, $e_1 \cdot e_2$ and $e_1 + e_2$ are also terms.

As the initial (undefined) relation, Poretskiy selects the relation of equality ($=$). The inclusion relation ($\leqslant$) is defined by use of the equality relation. All the formulas of the system we present will be numbered in succession; the letter A will be used to denote an axiom; the letter π, a rule of inference (the rules of inference are the formulas containing the symbols for implication $\rightarrow$ and conjunction &). A formula that is dual to another is denoted by the same number followed by an asterisk.

1A. $e = e$	(reflexivity of equality: principle of identity),
2π. $(e = e_1) \rightarrow (e_1 = e)$	(symmetry of equality),
3π. $((e = e_1) \mathbin{\&} (e_1 = e_2)) \rightarrow (e = e_2)$	(transitivity of equality),
4A. $e \cdot e = e$ 4*A. $e + e = e$	idempotence of addition and multiplication (disjunction and conjunction),
5A. $e \cdot e_1 = e_1 \cdot e$ 5*A. $e + e_1 = e_1 + e$	commutativity of addition and multiplication (disjunction and conjunction),
6A. $(e \cdot e_1) \cdot e_2 = e(e_1 \cdot e_2)$ 6*A. $(e + e_1) + e_2 = e + (e_1 + e_2)$	associativity of disjunction and conjunction,
7A. $e \cdot (e + e_1) = e$ 7*A. $e + ee_1 = e$	principles of absorption,

8A. $e \leqslant e_{1\,Df} : \equiv : e = ee_1$,†
8*A. $e_1 \geqslant e_{Df} : \equiv : e_1 = e_1 + e$,‡
9π. $(e = e_1) \rightarrow (e_2 \cdot e = e_2 \cdot e_1)$,
9*π. $(e = e_1) \rightarrow (e_2 + e = e_2 + e_1)$,
10A. $e \cdot (e_1 + e_2) = ee_1 + ee_2$,
10*A. $e + e_1 \cdot e_2 = (e + e_1) \cdot (e + e_2)$ } distributivity,
11A. $e + e' = 1$,
11*A. $e \cdot e' = 0$,
12A. $e + 0 = e$,
12*A. $e \cdot 1 = e$.

We consider the above representations as schemata for axioms, that is, as statements in which the letters e_1, e_2, e denote simple or compound terms. We shall illustrate the use of the rules of inference by explicitly analyzing Relationship 3π, which can be rendered verbally as follows: "If we have proved the premises $e = e_1$ and $e_1 = e_2$, we have proved the conclusion $e = e_2$ (the property of transitivity)." In other words, given the premises $e = e_1$ and $e_1 = e_2$, we can apply Rule 3π to them and derive from them the statement $e = e_2$.

All of Poretskiy's theorems concerning logical equations can be obtained from the axioms presented by use of the indicated rules of inference. Let us prove, for example, one of the theorems from Poretskiy's theory of logical consequents (Reference 404, pp. 183–216): $(e = e \cdot e_1) \rightarrow (e = e(e_1 + x))$.

In Poretskiy's interpretation, this theorem denotes the following: if we have proved the equation $e = ee_1$ or, in other words, $e \leqslant e_1$ (on the strength of Axiom 8A), we have also proved the equation $e = e(e_1 + x)$ (or, in other words, $e \leqslant (e_1 + x)$).

† In the interpretation in the logic of classes, the relation $a \leqslant b$ corresponds to inclusion of class a in class b.
‡ The relation $\geqslant$ is the inverse of the relation $\leqslant$. The expression $a \geqslant b$ can be read as "a contains (includes) b." Poretskiy expresses inclusion by means of the equality relation. Careful evaluation of the system of axioms 1A–12*A and the subsequent procedure leads us to the conclusion that Poretisky has actually described a structure (in the sense of contemporary abstract algebra) in terms of the undefined relation of equality. By proposing distributivity, Poretskiy showed that he considered the logic of classes to be a distributive structure: and by considering the Boolean operation of subtraction unnecessary, he refused, so to speak, to treat the logic of classes as a subtractive structure.

Proof:

1. $e = ee_1$ (given);
2. $e + ex = ee_1 + ex$ (by Rule $9^*\pi$ with e_2 equal to $e \cdot x$);
3. $e = e + ex$ (by 7*A and 2π).

From 3 and 2, using the rule of inference (3π), we obtain

4. $e = e \cdot e_1 + e \cdot x$;

then, by applying Formulas 10A and 2π, we have

5. $e = e \cdot (e + x)$, which was to be proved.

We note that, from the system of axioms and rules 1A through 12*A, it is possible to obtain the rule for substitution of equals for equals; we will henceforth make use of this rule in proving theorems in Poretskiy's theory. We shall also make use of the so-called duality principle, which follows from the preceding system of axioms (as can be easily shown). Nor will we prove here the theorems $0 \leqslant a$ for any class a and (dually) $a \leqslant 1$ for any class a, which we will employ. Their proof presents no difficulty.

Let us now move on to a description of the fundamental concepts of Poretskiy's theory. To solve a given logical equation "means to deduce from it all or some of its logical consequents" (Reference 401, pp. 11–12). Solution of an equation can be complete or partial, depending on whether all or only some of the consequents of the given equation are actually found. In speaking of "all the consequents," Poretskiy has in mind the deduction of those consequents of the given equation from which the equation itself can be obtained as a consequent. In other words, the complete solution of a given equation is that system of consequents which is equivalent to the equation itself. Thus, for example, one of the particular solutions of the equation

$$1 = ab + a'c \tag{1}$$

is

$$a = ab \tag{2}$$

but the complete solution of Equation 1 consists of the system of equations

$$a = ab \qquad \text{and} \qquad a' = a' \cdot c. \tag{3}$$

To see this, we multiply both sides of Equation 1 by a, obtaining $a = ab$. Indeed, we have

$$a \cdot 1 = a \cdot a \cdot b + a \cdot a' \cdot c, \qquad a = ab + 0, \qquad a = ab.$$

In the same way, multiplying both sides of Equation 1 by a', we obtain $a' \cdot 1 = a' \cdot a \cdot b + a' \cdot a' \cdot c$, or

$$a' = a' \cdot c. \tag{4}$$

Thus, Equations 2 and 4 are both consequents of Equation 1. It is easy to prove that Equation 1, in turn, is a consequent of the system of equations (3). In fact, multiplying Equations 2 and 4 termwise, we have

$$a + a' = ab + a' \cdot c \qquad \text{or} \qquad 1 = ab + a' \cdot c.$$

Thus, system 3 is the complete solution of Equation 1. On the other hand, it is clear that neither Equation 2 nor 4 singly is equivalent to Equation 1. To prove this, it is sufficient to set $b = 1$, a different from 1, and c equal to zero; then Equation 1 becomes untrue: we have that $1 = a \cdot 1 + a' \cdot 0$, that is, $1 = a$, which contradicts our condition that a is different from 1. Analogously, we can prove that Equation 1 does not follow from Equation 4.

Moving on to the representation of propositions in Poretskiy's system, we note first of all that, in order to understand Poretskiy's logical calculus, the following theorem is essential.

Theorem 1: For every equation $e_1 = e_2$, where $e_1 = \Phi(a, b, c, \ldots)$, $e_2 = \varphi(a, b, c, \ldots)$, there exist among its equivalent representations the following two forms:

$$e_1 \cdot e_2' + e_1' \cdot e_2 = 0,$$
$$e_1 \cdot e_2 + e_1' \cdot e_2' = 1.$$

We shall prove that $(e_1 = e_2) \rightleftarrows (e_1 e_2' + e_1' e_2 = 0)$, after which the question of the equation $e_1 \cdot e_2 + e_1' \cdot e_2' = 1$ will be solved through the duality principle.

We first prove* that $(e_1 = e_2) \leftrightarrows (e_1 e_2' + e_1' e_2 = 0)$. Thus, we

* In this proof we will not completely enumerate all of the axioms and rules we use; we will limit ourselves to those remarks from which the reader can easily reconstruct the entire course of the proof.

have the equation $e_1 = e_2$ (given). From this we obtain $e_1 \cdot e_2' = 0$ (by Rule 9π). From the original equation, we also obtain $e_1' \cdot e_2 = 0$ (by 9π and 5A).

Adding the preceding two equations termwise, we have

$$e_1 \cdot e_2' + e_1' \cdot e_2 = 0.$$

We shall now prove the converse: given

$$(e_1 \cdot e_2' + e_1' \cdot e_2 = 0) \rightarrow (e_1 = e_2), \qquad e_1 \cdot e_2' + e_1' \cdot e_2 = 0,$$

then

$$\left.\begin{aligned} e_1 \cdot e_2 &= 0 \qquad \text{(by } 9\pi \text{ and 12A),} \\ e_1' \cdot e_2 &= 0 \qquad \text{(by } 9\pi \text{ and 12A).} \end{aligned}\right\} \qquad \text{(I)}$$

To transform the two preceding equations into inclusions, we note that, from the system of axioms and Rules 1A through 12*A, we can derive the following relationship:

$$(e_1 \leqslant e_2) \rightleftarrows (e_1 \cdot e_2' = 0). \qquad \text{(II)}$$

To prove this, it is sufficient to make use of the definition of the relation $\leqslant$ in Rule 8A; multiplying both terms on the right-hand side of equivalence 8A by e_2', we obtain $(e_1 \leqslant e_2) \rightarrow (e_1 \cdot e_2' = 0)$. The implication in the other direction is obtained as follows: $e_1 = e_1(e_2 + e_2')$, but $e_1 \cdot e_2' = 0$, so that $e_1 = e_1 \cdot e_2$, that is, $e_1 \leqslant e_2$. Thus, from the system of equations I, taking into account Relationship II, we can conclude that $(e_1 \leqslant e_2)$ & $(e_2 \leqslant e_1)$. Consequently, we have that $e_1 = e_2$, which was to be proved.

It is easy to find a simple topological model for Theorem 1. Let $e_1 = \Phi(a) = a$; $e_2 = \varphi(b) = b$, where a and b denote classes (volumes of concepts), which in accordance with the conventions in logic, we will interpret as Euler circles. As a result, we have the diagram shown in Figure 26. From this diagram it is clear that, if $a = b$, the classes $a \cdot b'$ and $a' \cdot b$ are empty; it is likewise clear that the union of the classes $a \cdot b$ and $a' \cdot b'$ in this case coincides with the universal class 1.

However, the analytic proof of Theorem 1 presented is completely independent of any geometric interpretation of its conditions. Poretskiy calls the expression $e_1 e_2' + e_1' e_2$ in Theorem 1 the logical zero of the equation $e_1 = e_2$; we will denote this by

$\boxed{0}\,(e_1, e_2)$ or simply by $\boxed{0}$. Poretskiy calls the expression $e_1 \cdot e_2 + e_1' \cdot e_2'$ in Theorem 1 the logical unit of the equation $e_1 = e_2$; we will denote this by $\boxed{1}\,(e_1, e_2)$ or simply by $\boxed{1}$. Poretskiy also defines the "universal logical unit" as

$$1_{Df} : \equiv : (e_1 + e_1') \cdot (e_2 + e_2') \cdots (e_{n-1} + e_{n-1}) \cdot (e_n + e_n')$$

and the "universal logical zero" as

$$0_{Df} : \equiv : e \cdot e_1' + e_2 \cdot e_2' + \cdots + e_{n-1} \cdot e_{n-1}' + e_n \cdot e_n',$$

that is, the universal and empty classes, in contemporary terminology.

Another essential concept in Poretskiy's logical theory is the concept of a "form" of a given logical equation. A *form* of a given equation r is an equivalent representation r^*, such that r implies r^* and r^* implies r. For example, the equation $a = ab$ has the following forms: $ab' = 0$; $b = b + a$; $a' + b = 1$, and so on (see Figure 26).

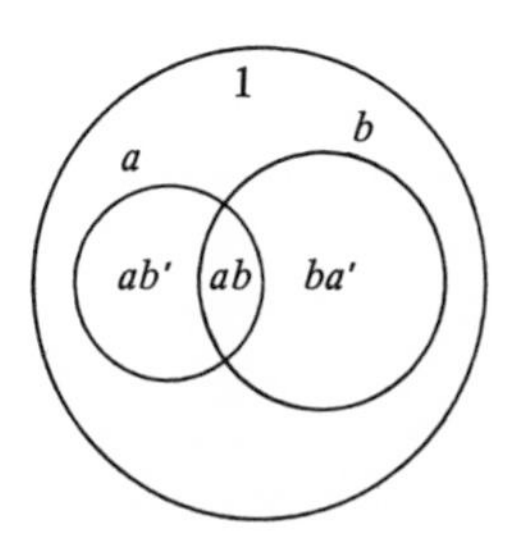

Figure 26

We should note that there are many logical expressions whose effective investigation can be greatly facilitated by application of Poretskiy's theory of forms. For example, suppose we apply Poretskiy's method to the system of equations $a = b = c$, whose logical zeros would be the following: (1) $a \cdot b' + a' \cdot b$ for $a = b$; (2) $bc' + b'c$ for $b = c$; (3) $ac' + a'c$ for $a = c$. Then we obtain the logical zero for the system in the form

$$ab' + a'b + bc' + b'c + ac' + a'c = 0.$$

The number of forms of a given equation will be finite only if we agree that these forms shall not contain letters which are not already a part of the original equation. Poretskiy himself expressed the theorem on forms by the following relationship:

$$(A = B) \rightleftarrows (x = x(AB + A'B') + x'(AB' + A'B)), \quad (1)$$

where the symbol $\rightleftarrows$ indicates the relation of equivalence; where

$A = \Phi(a, b, c, \ldots), \quad B = \varphi(a, b, c, \ldots);$ and where $AB' + A'B$

is the logical zero of the equation $A = B$, and $AB + A'B'$ is its logical unit; the symbol x represents an arbitrary term or a function of terms (in the sense of the algebra of logic), taken from the given equation $A = B$. We will not carry out a rigorous proof of this theorem but will limit ourselves to a simple verification of Relationship 1. We have

$$x = (AA' + A'A)x' + (AA + A'A')x = 0 \cdot x' + 1 \cdot x = x.$$

Since for every class K there exists the complement K' and, at the same time, every class K can be represented either in the form $\prod\sum$ (as a product of sums) or in the form $\sum\prod$ (as a sum of products), the number of all possible different classes which can be formed from n elementary classes is exactly equal to 2^{2^n}. Therefore, the arbitrary variable term x in Relationship 1 can take only 2^{2^n} different values. Because of the symmetry of the equivalence relation, the number of its forms is actually limited to $2^{2^n - 1}$ (provided, of course, that x enters into $A = B$ or, more exactly, that x is a function depending on the same variables on which A and B depend or on only certain of these variables).

We note in passing that Poretskiy's law of forms is still used by logicians today. Thus, in a relatively recent book by O. Becker, *Einführung in die Logistik*, we read: "Poretskiy's law of forms leads to important relationships for logical inequalities" (Reference 416, p. 21).

In the propositional calculus, Poretskiy's law of forms makes it possible to actually write out, for any given equivalence $a = b$, an expression that will be equivalent to the original one and will have a previously specified expression t on one side of its formula. Thus, Poretskiy's theorem makes possible solution of the equation $(a = b) = (x = t)$ with respect to x. This equation is solvable for any a, b, and t. Let us solve it for x (introducing, below, the $+$ sign for Boolean disjunction). We obtain, successively

$$\begin{aligned} (ab' + a'b)' &= (xt' + x't)'; \\ ab' + a'b &= xt' + x't; \\ a + b &= x + t; \\ x &= a + b + t \end{aligned}$$

or

$$x = (ab' + a'b) + t$$

or

$$x = (ab' + a'b)t' + (ab + a'b')t.$$

The final result is correct since

$$(a = b) = \{t = [(ab' + a'b)t' + (ab + a'b')t]\};$$

that is, we have obtained the theorem of forms. Thus, the theorem of forms can be derived on the basis of solution of Boolean equations with strict disjunction.

The natural generalization of the theorem of forms of logical equations is the theorem for their combination or, in another formulation, the theorem for replacing a system of premises by a single premise equivalent to the entire system. Poretskiy proved that a system of equations

$$A_1 = B_1, A_2 = B_2, \ldots, A_p = B_p,$$

where $A_i = \Phi_i(a, b, c, \ldots)$, $B_q = \varphi_q(a, b, c, \ldots)$, is equivalent to the single equation

$$0 = \sum_{n=1}^{n=p} (A_n \cdot B_n' + A_n' \cdot B_n),$$

where $\sum$ is the summation sign with respect to n. By the duality principle, the same system of equations is equivalent to the expression

$$1 = \prod_{n=1}^{n=p} (A_n \cdot B_n + A'_n \cdot B'_n),$$

where $\prod$ is the symbol for the product taken over n. We can use the symbols $\boxed{1}^{(n)}$ and $\boxed{0}^{(n)}$, respectively, as abbreviations for the expressions $A_n \cdot B_n + A'_n \cdot B'_n$ and $A_n \cdot B'_n + A'_n \cdot B_n$.

Poretskiy's combination theorem, in its most general form, can be stated as follows: In order to replace a given system of premises by a single premise equivalent to the system, it is sufficient to write all of the premise equations in the form of definitions of an arbitrary class x taken from these equations and then to write that x lies in the intersection of all the $\boxed{1}^{(n)}$ and contains within itself the union of all the $\boxed{0}^{(n)}$. This formulation clearly demonstrates that the combination theorem is actually a generalization of the theorem of forms.

Poretskiy's algorithm for eliminating nonessential variables is also very interesting. It is based on a theorem discovered by Poretskiy,† according to which we have the equivalence

$$(Ax + Bx') \rightleftarrows (Ax + Bx' + AB), \tag{2}$$

which can also be written in its dual form:

$$(Ax + Bx') \rightleftarrows ((Ax + Bx') \cdot (A + B)). \tag{2*}$$

Application of this theorem of Poretskiy's greatly simplifies the methods of elimination (exclusion) of variables, deduction of consequents from given premises, finding of premises for given facts, as well as methods of solution of so-called Boolean equations by means of their reduction to the simplest possible canonical form. Poretskiy's reduced normal form evidently can also be used

† See Reference 405, chapter LIV, p. 159. By expanding the third disjunctive term on the right-hand side of Equivalence 2 with respect to x, we note that both terms of this expansion (ABx and ABx') are absorbed by the terms Ax and Bx', respectively. This proves Poretskiy's theorem.

for simplifying structural formulas of relay–contact networks (see p. 241f. of this book).

Poretskiy's algorithm for eliminating nonessential variables takes the following form.† The most general form for a logical polynomial is the following: $A \cdot x + B \cdot x' + C$, where A, B, and C are logical functions that do not depend on the argument x. It is easy to see that the expression $AB + C$ is a subclass of the expression $Ax + Bx' + C$ and the expression $A + B + C$ is a "superclass" of the expression $Ax + Bx' + C$ (that is, it includes the latter as a subclass). Symbolically:

$$(AB + C) \leqslant (Ax + Bx' + C) \leqslant (A + B + C), \qquad (3)$$

where the symbol $\leqslant$ denotes the inclusion relation. Since the relation $\leqslant$ (in the sense of formal inference) has all the properties of the symbol $\rightarrow$ (implication) in the propositional calculus, Expression 3 can easily be transformed into the dual relationship

$$(A + B + C)' \leqslant (Ax + Bx' + C)' \leqslant (AB + C)'. \qquad (3^*)$$

Relationships 3 and 3* express the elimination theorem. Therefore, in this theorem, we can distinguish two cases: the first, in which the resultant is one of the consequents of the polynomial to

† Reference 405, vol. 10, no. 2. The second theorem for elimination, pp. 160–168. Poretskiy applies the theory of forms of logical equations, first of all, to the problem of excluding classes from given logical equations (the problem of elimination). Poretskiy's rule for elimination can be symbolically represented as

$$(e_1 = e_2) \rightarrow [\boxed{1}\,(a, a', b, b', \ldots, 1, 1, 1, 1, \ldots) = 1],$$

where $\boxed{1} = e_1 \cdot e_2 + e_1' \cdot e_2' = \boxed{1}\,(a, a', b, b', \ldots, k, k', f, f', \ldots)$.

In other words, "in order to exclude from equation r, written in unit form $\boxed{1}\,(a, b, c, \ldots) = 1$, a given n classes, it is sufficient to replace all the excluded classes and their negations by 1" (see Reference 401 and Reference 316, p. 88). It must be emphasized that, although the unit representation is not unique, the elimination theorem is valid for any such representation. In the preceding citation, we have made the following changes to keep our notation uniform: we have replaced Poretskiy's symbol M with our equivalent symbol $\boxed{1}$, and instead of "the equation $A = B$" we have written "the equation r."

which the elimination theorem has been applied; the second, in which the resultant is one of the "reasons" for the given logical polynomial. The resultant that is a consequent of the polynomial is obtained from the relationships

$$(Ax + Bx' + C) \leqslant (A + B + C), \qquad (3^{**})$$

$$(Ax + Bx' + C)' \leqslant (AB + C)', \qquad (3^{***})$$

the resultant that is a reason for the polynomial, from the relationships

$$(AB + C) \leqslant (Ax + Bx' + C), \qquad (3^{\text{IV}})$$

$$(A + B + C)' \leqslant (Ax + Bx' + C)'. \qquad (3^{\text{V}})$$

Let us consider an example of the application of Poretskiy's elimination theorem. Let us eliminate the letters b and d from the equation $abc'd + a'b'cd' = 1$ to obtain a resultant that is a consequent of the given equation. The given equation (only in the propositional calculus) is equivalent to the expression $abc'd + a'b'cd'$. To this latter expression, we apply the property reflected in Relationship 3**, and we obtain $a \cdot 1 \cdot c' \cdot 1 + a' \cdot 1 \cdot c \cdot 1$, which is equivalent to $ac' + a'c$ or $a = c'$.

Poretskiy also proposed a method for constructing a special table for logical equations, from which it is possible to compute the consequents of, as well as the reasons for, a given equation. First of all, the given logical equation is reduced to the form $\boxed{0} = 0$, where $\boxed{0}$ is expanded with respect to all the variables which enter into the given equation. Let m be the number of subclasses of this expanded expression. Then the table constructed for the given equation will consist of 2^m rows and $2^{2^n - m}$ columns. In the upper left-hand box, formed by the intersection of the first row and the first column of the table, we place a zero. In the first column of the table, beginning with zero, we place an additional $2^m - 1$ expressions, each of which is a subclass of the expansion $\boxed{0}$. The first row consists of 2^p terms, which are elements of the expansion $\overline{\boxed{0}}$ and all possible combinations of these elements (that is, they are subclasses of the expansion $\boxed{1}$ of the given equation). Beginning with zero, row one ends with the expression for expansion $\boxed{1}$ of the given equation. By construction, every class x_i that occupies a separate box in the table is equal to the union $x_{i_1} + x_{i_2}$, where x_{i_1} is the class located at the head of the column in which x_i stands, and x_{i_2} is the class located at the beginning of the row in which x_i stands. Then, on the strength of the given logical equation, every class in the table is equal to all the other classes located in the same column it is in. Let us consider as an example the construction of the table of consequents for the equation $a = b$. We have $\boxed{0} = \bar{a}b + a\bar{b}$. The subclasses of $\boxed{0}$ are the following: $0, \bar{a}b, a\bar{b}, a\bar{b} + \bar{a}b$. They form the first column of the table. The first row of the table (from left to right) consists of the classes $0, ab, \bar{a}\bar{b}, ab + \bar{a}\bar{b}$. As a result, we obtain the table shown in Figure 27. The more interesting con-

0	ab	$\bar{a}\bar{b}$	$ab + \bar{a}\bar{b}$
$a\bar{b}$	a	$\bar{b}$	$a + \bar{b}$
$\bar{a}b$	b	$\bar{a}$	$\bar{a} + b$
$a\bar{b} + \bar{a}b$	$a + b$	$\bar{a} + \bar{b}$	1

Figure 27

sequents of the equation $a = b$, computed from this table, are $0 = a\bar{b}$; $\bar{a}\bar{b} = \bar{b}$; $a\bar{b} + \bar{a}b = a\bar{b}$; $a + \bar{b} = \bar{a} + b$; $0 = \bar{a}b$; $\bar{a}\bar{b} = \bar{a}$; $a\bar{b} + \bar{a}b = \bar{a}b$; $ab = a$; $ab + \bar{a}\bar{b} = a + \bar{b}$; $a + b = b$; $1 = \bar{a} + b$; $ab = b$; $ab + \bar{a}\bar{b} = \bar{a} + b$; $a + b = a$; $1 = a + \bar{b}$.

The table of consequents can also be used for obtaining all the reasons for a given logical equation. Due to the dual relationships we have already discussed, the table of reasons for the equation $a = b$ is at the same time the table of consequents for the opposite equation $a = \bar{b}$. Suppose, for example, that we wish to find the reasons for the equation $a = b$. First we construct the table of consequents for the opposite equation $a = \bar{b}$. This table takes the form shown in Figure 28.

Now, in order to deduce from this table the reasons for the equation $a = b$ (which is opposite to the equation $a = \bar{b}$ used in constructing the table) instead of the consequents of the equation

0	$a\bar{b}$	$\bar{a}b$	$a\bar{b} + \bar{a}b$
ab	a	b	$a + b$
$\bar{a}\bar{b}$	$\bar{b}$	$\bar{a}$	$\bar{a} + \bar{b}$
$ab + \bar{a}\bar{b}$	$a + \bar{b}$	$\bar{a} + b$	1

Figure 28

$a = \bar{b}$, it is sufficient to equate the complement of each class occupying a separate box in the table with each of the other classes located in the same column. It is easy to see that in the case of sufficiently complicated logical problems, application of Poretskiy's table becomes very difficult in practice; in such cases, we can use Poretskiy's so-called simplified canonical form, which, although it does not directly contain all of the consequents, allows us to obtain them by means of an immediate "syllogistic" deduction.

One of the most interesting parts of Poretskiy's logical theory, from the point of view of the contemporary algebra of logic, is this treatment of the deduction of consequents from given premises. What is interesting is that in the solution of this problem he did not limit himself to the use of completely conjunctive (disjunctive, dually) normal forms for logical expressions as, for example, did D. Hilbert and W. Ackermann (Reference 360).

Poretskiy's device was further expanded by A. Blake (Reference 397). Therefore, henceforth we will present Poretskiy's ideas concerning canonical forms for logical expressions in Blake's terminology. To do this, we must define his basic concepts: formal inclusion and syllogistic polynomial, both introduced under the clear influence of Poretskiy's logical system. If every term of the polynomial A is included in at least one term of the polynomial B (inclusion is here understood in the sense of the relation $\leqslant$), we will say that polynomial A is included in polynomial B formally.

If polynomial A is included in polynomial B, but not formally, this fact will necessarily be represented as $A \subset B$. For example, the polynomial $a'c$ is not formally included in the polynomial $a'b + b' \cdot c$. A syllogistic polynomial (which will be denoted by p^{syl}) is defined as a polynomial having the property that every polynomial included in it is included formally. In other words, there does not exist any polynomial p which is included in p^{syl} nonformally. Thus, a syllogistic polynomial very simply characterizes the class of all polynomials included in it. Before presenting an example of a syllogistic polynomial, we will describe the device, which originated with Poretskiy, for obtaining the syllogistic polynomial for a given logical expression (Reference 405).

We will call two letters "conjugate" if they are related to each other by mutual negation. For example, the letters a and a', a'' and a''' are conjugate. We will call two monomials conjugate provided that they differ from each other, in that one of them contains some letter c, whereas the other contains its negation c'; in addition, for the two monomials to be conjugate, we will require that there be no other letter f for which this is also true. For example, the monomials $abcd$ and $ae'f'gd'$ are conjugate. However, the monomials ab and $a'b'$ are not conjugate since they contain not one but two pairs of conjugate letters.

Following Blake's procedure, we introduce the operation of conjugation of conjugate monomials. This operation consists of taking the product of all the letters in the conjugate monomials, with the exception of the single pair of conjugate letters themselves.

We will denote this operation by using the symbol Δ in front of the conjugate monomials, and we will join the result to the initial expression by using the symbol $\stackrel{\circ}{=}$:

$$\Delta(Ac + BC') \stackrel{\circ}{=} AB.$$

In fact, it is easy to prove that $Ac + Bc' + AB = Ac + Bc'$; we need only use the fact that $AB = AB(c + c')$. We then obtain

$Ac + Bc' + AB = Ac + Bc' + ABc + ABc'$. But ABc is absorbed by the term Ac, and ABc' is absorbed by the term Bc', that is, we have $Ac + Bc' + AB = Ac + Bc'$.

Thus, we have proved that, if we carry out the operation of conjugation with respect to every pair of conjugate terms of a given polynomial and add the results of all these conjugations to the original polynomial, the new polynomial will still be equivalent to the original one.

Let us now suppose that a given logical expression e is reduced to an equivalent polynomial, which we will denote by p_e. As Blake proved, p_e can be uniquely reduced to an equivalent syllogistic polynomial p_e^{syl} by conjugation of all the conjugate monomials and use of the rule for absorption: $e + e \cdot e_1 = e$ (Axiom 7*A).

For convenience, we can perform the operation of conjugation first over all two-letter terms, then over two-letter and three-letter terms, then over three-letter terms, and so on. After carrying out all possible conjugations and adding all the results to the polynomial as new terms, we carry out all possible absorptions. To illustrate this device, let us find, for example, the syllogistic polynomial for the expression

$$a'b'c + a'b'de + abd'de' + acd'e' + b + d + c'. \tag{1}$$

First of all, we can make the very simple observation that the third term is absorbed by the fifth; the second, by the sixth. Thus, Expression 1 can be reduced to the expression

$$a'b'c + acd'e' + b + d + c'. \tag{1*}$$

Let us now proceed to the conjugation of conjugate monomials. The monomial $a'b'c$ is conjugate to the monomial b. Thus we form the new term $a'c$, which we add to Expression 1*; but $a'c$ absorbs the term $a'b'c$. Conjugation of $acd'e'$ and c' yields the new term $ad'e'$. Thus, Expression 1* is reduced to the expression

$$a'c + ad'e' + b + d + c'. \tag{1**}$$

Conjugating $a'c$ and $ad'e'$, we obtain $cd'e'$. This latter term, conjugated with d, yields the new term ce'. Conjugating the term

ce' with c', we obtain the term e', which when added to Expression 1** absorbs the term $ad'e'$ as well as the newly formed term ce'. The term $a'c$, when conjugated with c', yields the new term a', which absorbs $a'c$. As a result, we obtain the simplified canonical form — the syllogistic polynomial

$$a' + b + d + e' + c', \tag{2}$$

which is the simplest representation of logical expression 1.

The syllogistic polynomial can be used to deduce consequents from a given logical expression. If the equation $B = 0$ is a logical consequent of the equation $A = 0$, then the polynomial p_B, equal to B, is included in the polynomial p_A^{syl}, equal to A.

Once we have reduced the polynomial p_A to the form p_A^{syl}, then p_B is included in p_A^{syl} formally. Therefore, every consequent of the logical equation $A = 0$ is included formally in the polynomial p_A^{syl}. Thus, by simply examining the syllogistic polynomial carefully, we can survey all the polynomials included in it, that is, all the nontrivial logical consequents of the equation $A = 0$.

As an illustration, let us consider the following example. The equation $p_A = (a'b + b'c = 0)$ corresponds to the system of equations $a' \cdot b = 0$ and $b' \cdot c = 0$, that is, to the system of premises $a \leqslant b$ and $b \leqslant c$ ("all a are b," and "all b are c"). The equation $p_B = (a' \cdot c = 0)$ corresponds to the conclusion $a \leqslant c$ ("all a are c"). However, we can easily see that $a' \cdot c$ is included in $a'b + b'c$ nonformally. But, once we have reduced p_A by conjugation, we obtain the form $p_A^{\text{syl}} = a'b + b'c + a'c$. Now $a'c$, that is, p_B, is included in p_A^{syl} formally.

Evidently, the method presented above for the reduction of a polynomial p_e to an equivalent polynomial p_e^{syl} also finds application in the theory of relay-contact networks. In Reference 417 it is proved that the laws of the algebra of logic also hold for the combination of various elements in relay-contact networks, where the logical product of statements corresponds to a series connection of contacts, and the logical sum + corresponds to the connection of contacts in parallel; logical negation ′ corresponds to opening closed contacts and closing open contacts.

The method just presented for obtaining the canonical form of a logical expression can be applied to the problem of simplifying the structural formulas of relay-contact networks. For example, Expression 1*, $a'b'c + acd'e' + b + d + c'$, corresponds to a network diagram of the form shown in Figure 29.

Poretskiy's simplified form for this expression, namely the Expression 2, $a' + b + d + e' + c'$, corresponds to the much simpler network diagram shown in Figure 30.

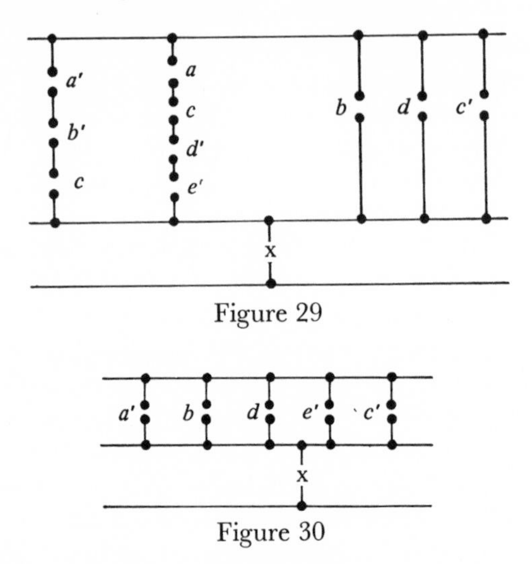

Figure 29

Figure 30

Poretskiy considered logical inequalities (References 414, 406, and 407). He analyzed them (and, therefore, also the unified theory of logical equalities and inequalities) from the point of view that when we are limited to a calculus with only one relationship of equivalence, it is impossible to express (in this calculus) even all the problems of classical logic. In particular, it is not possible to represent the partial statements of the Aristotelian formalism. However, it is also obvious that the problem of introducing an inequality relation causes certain difficulties. First of all, there arises the question of how we should understand the expression "x is not equal to y." Should we understand it in the sense of the assertion "it is not true that class x is equivalent to class y," thus introducing truth and falsity; or should we understand it as a designation of a particular subject, defined by the expression $x = \bar{y}$, where $\bar{y}$ is construed as meaning "not-y"? This latter understanding is expedient when the calculus of logical inequalities is to be interpreted in the domain of the propositional logic, for in this case we are dealing with a subject domain consisting of one single element, which can assume a set of values consisting only of truth and falsity. Of course, such an interpretation of the inequality relation is inadmissible if we are dealing with that part of the calculus of classes which is not isomorphic to the propositional calculus (that is, the so-called combinatorial propositional calculus of single-place predicates). For these reasons, Poretskiy first constructed his calculus of logical inequalities without fixing the relationship between equivalence and inequality, that is, without introducing rules which would allow for conversion from formulas containing the symbol $=$ to formulas containing the symbol $\neq$, and vice versa. The rules for such conversions were introduced by Poretskiy only later, when he had moved on to the interpretation of the calculus of logical inequalities within the propositional logic. Let us now analyze briefly the results obtained by Poretskiy.

We will consider the calculus of logical inequalities in which the expression $x \neq y$ is understood in the sense of the assertion "the relationship $x = y$ is false." In this calculus, we can establish, for example, the following formulas:

$$((a + b = x) \ \& \ (b \neq x)) \rightarrow (a \neq 0); \tag{1}$$

$$((a = 0) \ \& \ (b \neq x)) \rightarrow (a + b \neq x); \tag{2}$$

$$((a + b) \neq 0) \ \& \ (a = 0)) \rightarrow (b \neq 0); \tag{3}$$

$$(a \neq 0) \equiv (x = a\bar{x} + \bar{a}x); \tag{4}$$

$$(a \neq b) \equiv (x \neq (ab + \bar{a}\bar{b})x + (a\bar{b} + \bar{a}b)\bar{x}); \tag{5}$$

$$(Ax + B\bar{x} \neq 0) \rightarrow (A + B \neq 0). \tag{6}$$

The truth of these expressions is easy to verify. As an example, we will verify Expressions 2 and 6. Suppose that in Expression 2 a is equal to zero and $a + b = x$; then, in this case, b would be equal to x, contradicting the second premise, $b \neq x$. Now, let us prove Expression 6. The relationship $(Ax + B\bar{x}) \leqslant (A + B)$ has been proved in the calculus of logical equalities. Since the antecedent in the preceding expression is different from zero (see the premise in Expression 6), then the equation $A + B = 0$ must be considered false; but this, by the definition of the symbol $\neq$, means that $A + B \neq 0$, which was to be proved.

Let us now touch on the problems Poretskiy investigated in his unified theory of logical equalities and inequalities (Reference 407). One of the fundamental theorems of this theory is presented here.

Theorem: The condition under which the equation $Ax + B\bar{x} = 0$ and the inequality $Cx + D\bar{x} \neq 0$ can be simultaneously satisfied is given by the conjunction of the equation $AB = 0$ and the inequality $C\bar{A} + \bar{B}D \neq 0$. In this case, the determination of x resulting from the original formulas (that is, the "solution" of the original formulas with respect to x) takes the form

$$x \neq (\bar{A}\bar{C} + A\bar{D})x + (BC + \bar{B}D)\bar{x}.$$

Proof: The equivalence $(Ax + B\bar{x} = 0) \equiv (x = \bar{A}x + \bar{B}x)$ has been proved in the calculus of logical equations. Substituting this value of x in the expression $Cx + D\bar{x} \neq 0$, we obtain

$$(\bar{A}C + AD)x + (BC + \bar{B}D)\bar{x} \neq 0.$$

Taking account of Relationship 6, just proved, we can conclude from the preceding formula that $\bar{A}C + AD + BC + \bar{B}D \neq 0$, which we will find convenient to group in the form

$$(\bar{A} + B)C + (\bar{A} + B)D \neq 0. \tag{7}$$

Since the equation $AB = 0$ is a necessary condition for the premise $Ax + B\bar{x} = 0$ to be possible, we have that

$$\bar{A} + B = (\bar{A} + B)\cdot(A + \bar{A}) = \bar{A}$$

and, on the other hand, $A + \bar{B} = (A + \bar{B})\cdot(B + \bar{B}) = \bar{B}$ (for $AB = 0$!). Inequality 7 can therefore be written in the form $\bar{A}C + \bar{B}D \neq 0$. Hence, the condition under which the given equality and inequality are simultaneously fulfilled is the conjunction $(AB = 0)$ & $(\bar{A}C + \bar{B}D \neq 0)$.

Let us now prove the second part of the theorem. On the strength of the equivalence $(Ax + B\bar{x} \neq 0) \equiv (x \neq \bar{A}x + B\bar{x})$, we have

$$(\bar{A}C + AD)x + (BC + \bar{B}D)\bar{x} \neq 0,$$

which is equivalent to $x \neq \overline{\bar{A}C + AD} \cdot x + (BC + \bar{B}D) \cdot x$, and from which we obtain $x \neq (\bar{A}\bar{C} + A\bar{D}) \cdot x + (BC + \bar{B}D) \cdot \bar{x}$. The theorem is proved.

Later, Poretskiy turned to the interpretation of his algorithm (for the unified theory of equalities and inequalities) in the propositional calculus. The letters $a, b, c, d, \ldots$ now do not denote predicates or classes but individual statements; these statements are considered not from the point of view of an analysis of their internal structure but as single indecomposable wholes. In accordance with this interpretation, metasymbols of the form e_j will be used as abbreviations for substantive remarks about the statements $a, b, c, \ldots$ The apparatus for the new calculus that arises in this case demands an expansion of the axiomatization developed earlier, and Poretskiy actually accomplishes this expansion. He introduces the following axioms for the propositional logic (the so-called principles of assertion and negation):

13A. $(e = 1) = e$,
13*A. $(e = 0) = \bar{e}$.

He then proves the theorem "If $e = 0$, then $\bar{e} = 1$, and if $e = 1$, then $\bar{e} = 0$" and one other important theorem which allows us to eliminate the symbol $\neq$ in the propositional calculus (Reference 407, p. 29, par. 106), namely, "The following equivalence holds (in the propositional logic): $a \neq b$ is equivalent to $a = \bar{b}$."

The proof can be carried out by a chain of equivalences:

$$(a \neq b) \equiv (a\bar{b} + \bar{a}b \neq 0) \equiv (a\bar{b} + \bar{a}b = 1) \equiv (a = \bar{b}).$$

Poretskiy dealt successfully with the problem of solvability in the calculus of propositions with inequality. Solution of the solvability problem in each specific case is accomplished by carrying out the following three steps in sequence:

1. the expression of all implications and nonimplications (that

is, relationships of the form $x \rightarrow y$ and $x \not\rightarrow y$) by use of equivalences and nonequivalences, respectively;

2. the reduction of the system of equations obtained in this way to unit form, and reduction of the inequalities to zero form (so that later they can easily be transformed into unit form);

3. the application of conjunction to the system of equivalences conjunctively linked, and of disjunction to the system of equivalences connected disjunctively; in the course of this process, Axiom 13A is used (to eliminate the unit representations). Finally, we remove brackets and simplify.

As an example of the application of Poretskiy's rules, we will check to see whether or not the following expression is identically true: $(a \not\rightarrow b)\ \&\ ((c \rightarrow \bar{d}) \vee (e \not\rightarrow \bar{b}))\ \&\ (\bar{c} \rightarrow d)$.

Transforming this expression by using Rules 1 through 3, we have

$$(a \neq ab)\ \&\ ((c = c\bar{d}) \vee (e \neq e\bar{b}))\ \&\ (\bar{c} = \bar{c}d),$$

$$(\bar{a} + b = 0)\ \&\ ((\bar{c} + \bar{d} = 1) \vee (\bar{e} + \bar{b} = 0))\ \&\ (c + d = 1),$$

from $(\overline{\bar{a} + b}) \cdot (\bar{c} + \bar{d} + eb) \cdot (c + d)$, or $a\bar{b} \cdot (c\bar{d} + \bar{c}d)$, which indicates that the given expression is not a tautology.

It should be noted here that all the theorems proved in the calculus of logical equations are still valid in the calculus of propositions with inequality. In particular, the law of forms of statements (Reference 407, p. 38, par. 107) still holds.

Poretskiy did not limit himself to the investigation of syllogistic deduction; he also considered nonsyllogistic forms. It is of great significance for traditional logic that, on the basis of the generalized logical theory of classes worked out by Poretskiy, the essentially analytic nature of deductive implication can be easily demonstrated. The analytic nature of deduction is proved if only by the fact that the decomposition of the premises into elements of volume (extent) contains all those elementary classes present in the decompositions of the consequent. This was clear, for example, in the analysis of the deduction of syllogistic consequents from the premises $a \leqslant b$ and $b \leqslant c$.

Poretskiy criticized the representation of deduction as merely a process for exclusion of the middle terms. According to Poretskiy, the deductive process, in general, consists not of an exclusion of the middle terms but of an exclusion of excess data. What Poretskiy means by this is that in passing from an equation $R = 0$ to one of its consequents, it is sufficient to eliminate from the left-hand side of the equation (represented as a polynomial in completely normal form) certain of its constituents. It should be emphasized that logic interested Poretskiy primarily as a method for obtaining definite results and not merely as an unorganized description of certain logical laws. Therefore, the algorithmic side of logic was his principal interest. Poretskiy understood that the algorithmic character of the methods he worked out would widen the possibilities for application of logic to other disciplines. In particular, Poretskiy applied his calculus of logical equations to the solution of certain problems in the mathematical theory of probability (Reference 402). Poretskiy stated the essential characteristics of a logical algorithm in the following way: first of all, this algorithm should contain a set of precise and unique prescriptions for operation at each step, that is, after completion of the preceding step; secondly, it should be of a more or less general nature; that is, it should relate not to one but to a whole class of homogeneous logical problems.

For certain branches of logic, such a definitive algorithm actually exists, for example, Poretskiy's algorithm in the theory of logical equations, which is absolutely valid in the propositional logic and in that part of the logic of classes which is isomorphic to the propositional calculus. However, as has become apparent in the course of further development of the science, not every homogeneous class of problems admits the existence of a general definitive algorithm. It has become clear that many classes of problems, for example, the identity problem in the theory of groups or the problem of finding a general method of solution for Diophantine equations, seem to be algorithmically unsolvable.

One of Poretskiy's Russian colleagues was the mathematician from Odessa, Evgeniy Leonidovich Bunitskiy (1874–1952). Bunitskiy was born in Simferopol. Having obtained his mathematical education from Odessa University, he worked for many years as a mathematics teacher in secondary schools in Odessa. He also participated actively in the sessions of the mathematics section of the Novorossiyskiy Society of Natural Scientists, one of whose members was I. V. Sleshinskiy, professor of mathematics at Odessa University and a great enthusiast and popularizer of the algebra of logic.*

In 1903 Bunitskiy was awarded the title of Master of Mathematics, and after 1904 he worked as an assistant professor and, later, as a full professor (1918–1922) at Odessa University. Bunitskiy died in Czechoslovakia in 1952, where he had worked for some time as a lecturer at Charles University, teaching mathematical analysis in the science department. His studies in the algebra of logic date exclusively from the Odessa period, specifically, from the years 1896–1899. He is the author of the following articles, published in the Russian journal *Vestnik Opytnoy Fiziki i Elementarnoy Matematiki* [Bulletin of experimental physics and elementary mathematics]: "Nekotorye prilozheniya matematicheskoy logiki k arifmetike" [Some applications of mathematical logic to arithmetic], nos. 247 and 248, 1896–1897; "Chislo elementov v logicheskom mnogochlene" [The number of elements in a logical polynomial], no. 249, 1897; "Nekotorye prilozheniya matematicheskoy logiki k teorii OND i NOK" [Some applications of mathematical logic to the theory of the greatest common divisor and the least common multiple], no. 274, 1899.

Bunitskiy was principally interested in the application of certain results of the algebra of logic to arithmetic and also in the question of determining the number of terms in a logical polynomial. We will present the results of his study of the so-called enumeration problem for a logical polynomial. If a polynomial is of the form $A \nabla B$, where there are no terms with a negation sign $(-)$ over them and where the symbol ∇ can stand for ordinary or Boolean union (that is, the union of given logical classes with the exclusion

* Already in 1893, Sleshinskiy had published an article, "Logicheskaya mashina Dzhevonsa" [Jevons' logical machine] (Reference 374), in an attempt to arouse interest in the algebra of logic among his colleagues. It was through his influence that Bunitskiy became acquainted with the algebra of logic.

of their common portions), the problem of "enumeration" for such a polynomial can be solved in the following way:

The formula

$$N(A \oplus B) = N(A) + N(B) \tag{1}$$

is obvious. Here the symbol $\oplus$ represents Boolean union (corresponding to the operation of strict disjunction in the propositional calculus); the numbers indicating how many different elements are contained in the classes $A, B, C, \ldots$ will be denoted by $N(A)$, $N(B)$, $N(C), \ldots$; the $+$ sign on the right-hand side of Equation 1 is the symbol for arithmetic addition. To distinguish it from the symbol for non-Boolean union, we will denote the latter by the symbol $\cup$. In the case of non-Boolean union we have

$$N(A \cup B) = N(A) + N(B) - N(AB), \tag{2}$$

where the $-$ on the right-hand side is the symbol for arithmetic subtraction. It is easy to extend Formula 2 to the case of union of three logical classes A, B, C. We obtain

$$\begin{aligned} N(A \cup B \cup C) &= N((A \cup B) \cup C) \\ &= N(A \cup B) + N(C) - N((A \cup B)C) \\ &= N(A) + N(B) + N(C) - N(AB) \\ &\quad - N(AC) - N(BC) + N(ABC). \end{aligned}$$

Thus, we have

$$\begin{aligned} N(A \cup B \cup C) = N(A) + N(B) + N(C) - N(AB) \\ - N(AC) - N(BC) + N(ABC). \end{aligned} \tag{3}$$

By using the method of mathematical induction, Bunitskiy finally arrives at the formula (Reference 333, p. 202)

$$\begin{aligned} N\left(\sum_{k=1}^{k=n} A_k\right) = \sum N({}_nP_1) - \sum N({}_nP_2) + \cdots \\ + (-1)^{m-1} \sum N({}_nP_m) + \cdots + (-1)^{n-1} \sum N({}_nP_n) \end{aligned} \tag{4}$$

where the symbol ${}_nP_m$ denotes the product of m classes P taken from among the classes $A_1, A_2, \ldots, A_{n-1}, A_n$, for $m > 1$. The expression ${}_nP_1$ denotes one of these n classes; the symbol $\sum$ in front of $N({}_nP_m)$ indicates that the sum of the expressions $N({}_nP_m)$ is to be taken over all possible ${}_nP_m$. Formula 4 allows us to calculate the number of pairwise nonequivalent elements in the compound logical class $A_1 \cup A_2 \cup \cdots \cup A_{n-1} \cup A_n$ if we know the number of elements in each of these classes individually. It also makes possible an analogous computation of the number of elements in all combinations of 2, 3, . . ., and finally $n - 1$ classes and computation of the number of elements in the product of all of these classes.

Then Bunitskiy proceeds to the problem of determining the number of elements in a logical polynomial in the case where the polynomial also includes negated terms. The formula

$$1 = A \cup \bar{A} \tag{5}$$

is obvious. From it, we can move on to the equation

$$N(1) = N(A \cup \bar{A}); \tag{6}$$

when we apply Equation 2 to the right-hand side of Equation 6 we obtain

$$N(1) = N(A) + N(\bar{A}) - N(A\bar{A}). \tag{7}$$

Using the obvious tautology $N(A\bar{A}) = N(0) = 0$, we transform Equation 7 to the form

$$N(1) = N(A) - N(\bar{A}), \tag{8}$$

from which it follows that

$$N(\bar{A}) = N(1) - N(A). \tag{9}$$

Bunitskiy further proved that Formula 9 can be generalized as follows:

$$N(P\bar{A}) = N(P) - N(PA), \tag{10}$$

where P is a positive class formed from nonnegative terms, or a product of positive classes. Later, he proved the equation

$$N(\bar{A}_1\bar{A}_2\bar{A}_3\cdots\bar{A}_{n-1}\bar{A}_n) = N(1) - N(A_1 \cup A_2 \cup \cdots \cup A_n), \tag{11}$$

which is a generalization of Equation 9. Combining Formulas 11 and 10, Bunitskiy obtained the general formula

$$N\left(P\prod_{i=1}^{i=n}\bar{A}^{(i)}\right) = N(P) - N(PA_1 \cup \cdots \cup PA_n), \qquad (12)$$

which is the solution to the problem in the case of the presence of negated terms in the polynomial. Relationships 4 and 12 are even used now for solving the problem of determining the number of elements in a logical polynomial — and thus this problem has proved soluble both for the case of a polynomial without negated terms and for the case which includes negated terms.

In Reference 425, Bunitskiy investigated symmetric functions in the algebra of logic; these differ from the usual symmetric functions found in mathematics in that the Boolean principle of absorption holds for them. Bunitskiy introduced the finite logical class A, whose elements are placed in correspondence with the prime numbers $\alpha_1, \alpha_2, \alpha_3, \alpha_4, \ldots. \alpha_{m-1}, \alpha_m$ (not necessarily different); the class A itself is placed in correspondence to the product of these numbers. This product, which is in one-to-one correspondence with the given logical class A, is denoted by $[A]$. In this way, finite classes can be considered as numbers, on the basis of their elements. The method of calculation makes it easy to determine the classes from the numbers. Given such an interpretation, "the logical sum of the classes $A, B, C, \ldots$ corresponds to the number which is equal to the least common multiple of the numbers $[A], [B], [C], \ldots,$ associated with the given classes; the logical product of the classes $A, B, C, \ldots$ corresponds to the number which is equal to the greatest common divisor of the numbers $[A], [B], [C], \ldots$" (Reference 425, p. 260).

But, in this case,

"Every logical operation on a series of classes $A, B, C, \ldots$ reduces to some proposition in the realm of the theory of the least common multiple and the greatest common divisor" (Reference 425, p. 260).

Bunitskiy also established the following facts: (1) every symmetric function in the algebra of logic can be represented as the sum of certain prime symmetric functions; (2) "the logical sum of two prime symmetric functions can easily be reduced to the one of them whose degree of homogeneity is lower" (Reference 425, p. 251). This latter fact, as is easy to see, is a result of the absorption principle. Furthermore, it turns out that every symmetric function in the algebra of logic can be reduced to a single prime symmetric function. This result is significant because it allows us to formulate the following arithmetic theorem:

"If we join the numbers $[A]$, $[B]$, $[C]$, ... in any way whatever (as long as it is done symmetrically with respect to these numbers), by a series of operations consisting of finding the greatest common divisor and the least common multiple, then, without changing the final result, this series of operations can be replaced in the following way: (1) we form groups of combinations of the given numbers $[A]$, $[B]$, $[C]$, ... taken K at a time; (2) we find the greatest common divisor of each group and then find the least common multiple of all the obtained divisors. The number K is the least number of different multipliers entering into the composition of the separate terms of that logical polynomial which is obtained if the given symmetric functions ... are replaced by the corresponding logical ones" (Reference 425, p. 251).

Bunitskiy applies these remarks to the simplification of a symmetric function in the algebra of logic of the form

$$\prod(A\alpha_1 + A\alpha_2 + \cdots + A\alpha_m),$$

where $\prod$ is the symbol for the product taken over all possible different sums of the form $A\alpha_1 + A\alpha_2 + \cdots + A\alpha_m$, and $\alpha_1, \ldots, \alpha_m$ are different numbers taken from the series of n consecutive numbers $1, 2, \ldots, n$. He proves that the equation

$$\prod(A\alpha_1 + A\alpha_2 + \cdots + A\alpha_m) = \sum A\alpha'_1 A\alpha'_2 \cdots A\alpha'_{n-m+1},$$

holds, where $\sum$ is the summation sign, and $\alpha'_1, \alpha'_2, \ldots, \alpha'_{n-m+1}$ take all possible different values among the numbers $1, 2, \ldots, n$. This result was given further arithmetic interpretation by Bunitskiy; however, we will not present this interpretation here since we are interested only in the general logical aspect of Bunitskiy's achievements. Furthermore, Bunitskiy's results in Reference 425 cannot be classified as an arithmetic model of the algebra of logic, in a strict sense. Rather, we should say (using contemporary terminology) that Bunitskiy's general theorem is a proposition for a structure in terms of the theory of symmetric functions; this proposition can then be translated into the corresponding arithmetic theorem. We should note that at the time of Bunitskiy's work in logic the concept of a structure had not yet been formulated, although it was implicitly contained in the works of Schröder (Reference 383). After 1900, Bunitskiy gradually turned away from active creative work in the field of the algebra of logic. However, Poretskiy continued to occupy himself very successfully with further studies in this field.

In Reference 405, Poretskiy introduced, in particular, the concept of a complex logical base. In the work *Teoriya logicheskikh neravenstv* [Theory of logical inequalities], Kazan, 1904, Poretskiy for the first time turned to the interpretation of his algorithm in the realm of the propositional calculus and constructed a logical theory that included elements of the combined calculus of propositions and classes.

The article "Théorie conjointe des égalités et des non-égalités logiques," published in Kazan in 1908 after the death of its author, considers questions of the relationships between the propositional logic and the combined calculus of propositions and one-place predicates.

Poretskiy also maintained a lively interest in the literature concerning mathematical logic published by his contemporaries. For example, he gave high praise to Louis Couturat's analysis of the logical system of the Italian mathematician Peano.

For literature on P. S. Poretskiy, see References 397–426.

3. The Problematics of the Algebra of Logic in the Works of C. S. Peirce, the Father of Semiotics

The American proponent of mathematical logic, Charles Sanders Peirce, made an essential contribution to the development of the logico-algebraic conception and became the founder of a new science — semiotics (the general theory of signs).

Peirce was born on September 10, 1839 in Cambridge, Massachusetts. His father was a professor at Harvard University. At the age of seven, Peirce began the study of chemistry and at twelve did independent work in a chemical laboratory. A year later, he diligently studied Whately's *Elements of Logic*, a book written from the point of view of inductive methodology. It apparently had a great influence on the formation of his own scientific views.

From 1861 to 1891 Peirce worked at a U.S. Coast Guard Geodesic Station. However, in addition to this principal occupation, he took on teaching duties in Cambridge and at Harvard. From 1869 to 1872, he was an assistant at the Harvard Observatory. He lectured on logic at Harvard from 1870 to 1871 and at Johns Hopkins University from 1879 to 1884.

Peirce was strongly attracted to the exact sciences (mathematics and logic) and also to philosophical problematics, under the influence of the ideas of Charles Darwin, Charles Lyell, and Duns Scotus. In particular, it was under the influence of the latter that Peirce reflected on the problem of universals. He gradually came to the conclusion that the Scholastic question as to the nature of universal concepts, was, in essence, of a linguistic character.

In 1887, Peirce settled in Milford, Pennsylvania. He died in poverty on April 19, 1914, forgotten by his friends and acquaintances. The high peaks he attained never gave him any refuge or shelter from the tiresome company of his creditors.

After Peirce's death, his manuscripts were given over to the custody of the Philosophy Department at Harvard University.

Peirce played an outstanding role in the history of American philosophical thought. During his lifetime he published no books on philosophy, and his ideas frequently were expressed only to the circle of students whom he taught. Articles on "pragmaticism" (as Peirce called his philosophical ideas, to differentiate them terminologically from James' pragmatism, of which he was critical) were only one aspect of Peirce's methodological work. In fact, the principal content of this work is the formulation and solution of problems in semiotics. He was also very much interested in the problem of induction and in the logical analysis of mathematics.

We should take note of the following printed works of Peirce in logic and philosophy: *On the Natural Classification of Arguments* (1865) — this concerns the types of argumentation used in proofs; *Logical Papers* (1867) — a collection of articles on various questions in formal logic; *On an Improvement in Boole's Calculus of Logic* (1867)

— an analysis of the difficulties connected with Boole's calculus; *Description of a Notation for the Logic of Relatives, Resulting from an Amplification of the Concepts of Boole's Calculus of Logic* (1870); *On the Algebra of Logic* (1880); *A Boolean Algebra with One Constant* (1880) — this contains an exact anticipation of the Sheffer stroke; *On the Logic of Number* (1881); *The Logic of Relatives* (1883) — here Peirce introduces quantifiers connecting relations; the symbol $\sum_i$ is used as an abbreviation for the expression "for some i," Π_j as an abbreviation for the phrase "for any j," where i and j are the names of relations; *On the Algebra of Logic* (1885) — here Peirce considers a logic without relations, an intensional logic of relations of the first degree, and an intensional logic of the second degree, and he introduces quantifiers with respect to predicates; *Logical Machines* (1888), and *The Logic of Relatives* (1897).

The multiple aspects of Peirce's methodological ideas makes it very difficult to classify them completely in an unambiguous fashion. First of all, we must note that Peirce's pragmaticism is not at all the same as the pragmatism of William James. James, being neither a mathematician nor a logician, was not able to fully grasp Peirce's ideas; he only greatly simplified and (in essence) inadequately adapted these ideas. Unfortunately, many investigators have not attached any significance to this latter fact and have wrongly depicted Peirce as undoubtedly a predecessor of James. As a matter of fact, Peirce's methodology is not subjectivist, and in certain points it even approaches conscious materialism (for example, on the question of the role of practical experience in the process of knowing objective reality).

Peirce recognized the existence of an external reality, independent of man; in fact, he even stated that the real can be defined as something whose properties are independent of what is thought about them (Reference 445, p. 42). Elsewhere he said that the basic hypothesis of the scientific method is "there exist real objects whose properties are completely independent of our opinion of them" (Reference 448, p. 243). Peirce defended scientific values against the attacks of the clerics and criticized the attempts of Cambridge intellectuals to reconcile the new science with theological fabrications.

Along with other members of the "Metaphysical Club," Peirce sharply disassociated himself from philosophical conceptions involving fatalistic notions as to primordial necessity in nature, which was treated as being exclusive of any randomness. Later, in his own theory of "tachism," he attempted to approach a correct understanding of the category of randomness.

Peirce defined pragmaticism as a general method for revealing the "real" meaning of any concept, doctrine, proposition, word, or any other *signs* (Reference 446, p. 38) [my italics — N.S.]. It is quite clear that James did not insert any such understanding in his conception of pragmatism. Even Peirce makes this assertion in stating that pragmatism does not pose as its problem the clarification of the meaning of all signs (Reference 448, p. 5). Moreover, pragmaticism distinguishes subjective and objective modality and advances its fundamental concepts according to a strictly logical plan, again treating problems that are absent in the pragmatism of James and Dewey. Thus, Peirce's pragmaticism essentially stands opposite to the pragmatism of James, at least with respect to objectivism and subjectivism.

According to Peirce, the thought of an idea can be exposed only by a study of those results to which it reduces in its practical use. In 1872 Peirce advanced the remarkable principle that in order to define the idea of an intellectual concept it is necessary to consider what practical consequences follow unavoidably from the truth of this concept. Then the sum of these consequences defines the idea of the concept (Reference 448, p. 255).

From a methodological point of view, this statement underscores the pre-eminence of practical experience as the truest criterion. From a logical point of view, it reduces to the demand that the reliable stock of concepts in the language of science include only verified statements.

Peirce was interested in the logical problem of the creation of an opinion and its reliability. He rejected the voluntaristic "solution" to this problem by condemning as defective the so-called "stability method," in which an opinion is considered immutably true as long as there are no counterexamples to it. He also rejected the "authoritative method" — that of imposing definite opinions on the individual by using the apparatus of the state machine. The only method of creating opinions that is worthy of our attention, said Peirce, is the scientific method. He asserted we must find a method that will aid our opinions to come forward not with a human but with some external constancy, in such a way that our thinking would have no influence (Reference 448, p. 25). Here the problem of finding an objective criterion for the validity of an opinion is formulated very explicitly.

In Peirce's pragmaticism, the concept of truth is given a logical formulation. The truth of a statement is considered from the point of view of the information expressed in it (the denotant of the concept is considered by Peirce to be the sender of information) as well as its verifiability.

Peirce can be considered the founder of semiotics and a pioneer in the field of logico-semantic investigations. He developed a science for studying any system of signs applied in human society. In particular, he attempted to study the language of science as a particular case of a system of signs. The subject matter of semiotic analysis, according to Peirce, comprises models of represented objects consisting of a finite set of elements and the relationships linking them.

Peirce defined a sign as an element x which, for the subject (as interpreter of the sign), takes the place of an element y (the denotant) according to an indication (or relation) R. According to Peirce, the most general subdivision of signs is the following: "icons, indices, and symbols" (Reference 450, p. 157).

The object of a sign is a thing. The basic function of a sign, in Peirce's opinion, is the quantizing of experiment or experience. His works already contained, in embryo, the division made by C. W. Morris of semiotics into pragmatics (concerning the relationship of the sign to its user), semantics (clarifying the relationship of the sign to the object denoted by it), and syntax (investigating the mutual relationships between signs). Peirce also analyzed the semiotics of the perception of signs, taking into account, in particular, the essential difference between perceived and communicated symbolic information. Within the context of his theory of signs, he considered the logical problem of meaning. Specifically, he described meaning as a relation between signs and the operations of a cognizant subject.

The point of departure for Peirce's investigations in logic was the work of George Boole. As early as 1870, Peirce used the term "Boolean logical algebra" to denote mathematical logic in contradistinction to traditional logic. Ten years later, he adopted A. Macfarlane's term "the algebra of logic."

In his investigation of the Boolean logical calculus, Peirce rigorously distinguished Boolean strict disjunction, which he considered an "arithmetic relation," from inclusive disjunction, which denoted the class of those things which were *a* or *b* or simultaneously *a* and *b* (Reference 451, p. 16).

By 1880, Peirce was already using the perfect disjunctive normal form as well as its dual, the perfect conjunctive normal form, in his calculus of classes (although admittedly with a notation that differs somewhat from that used today). These admitted simple syntactical interpretation for the calculus of propositions, which appears in Peirce's work in rather developed form. In particular, we cite the following laws of material implication formulated by Peirce in his calculus (the symbol $\prec$ corresponds to the compound "if . . . then"):

$$((x \prec y) \prec x) \prec x \qquad \text{(Peirce's law)}, \tag{1}$$

$$x \prec (y \prec x), \tag{2}$$

$$((x \prec y) \prec \alpha) \prec x, \tag{3}$$

where α denotes the identically false constant;

$$(x \prec (y \prec z)) \prec (y \prec (x \prec z)), \tag{4}$$

$$x \prec ((x \prec y) \prec y), \tag{5}$$

$$(x \prec y) \prec ((y \prec z) \prec (x \prec z)). \tag{6}$$

In addition to material implication, Peirce also advanced a substantial treatment of logical inference. Thus, in a certain sense, Peirce can be considered a predecessor of Lewis in the development of a theory of strict implication.

In 1885, Peirce exhibited a tabular procedure that presented a general method for treating the problem of solution in propositional logic. Peirce's truth table was constructed as follows: We draw a rhombus and then join the midpoints of the opposite sides by straight lines. Thus, the rhombus will be divided into four rectangles; in each rectangle we will write the corresponding truth

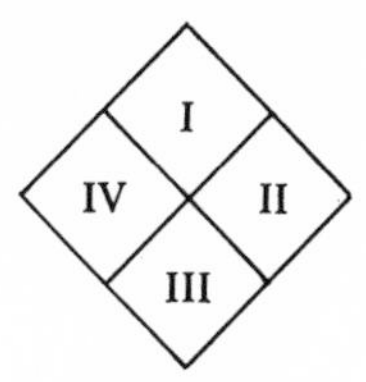

Figure 31

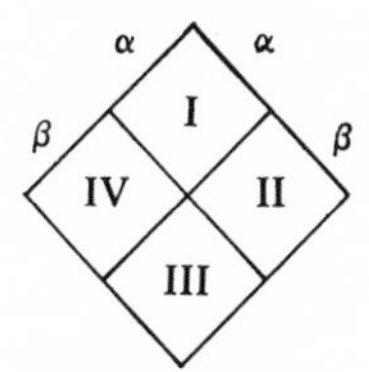

Figure 32

value of the function. Let us number the rectangles in clockwise order, using Roman numerals (see Figure 31). Suppose we choose only two truth values, α and β. The truth values of the arguments are recorded as shown in Figure 32.

We enter the truth values of the function in the table as follows: (I) $F(\alpha, \alpha)$; (II) $F(\alpha, \beta)$; (III) $F(\beta, \beta)$; (IV) $F(\beta, \alpha)$. (See Figure 33.) Suppose that α corresponds to false (symbolically, 0) and β to true (symbolically, 1); then the table for conjunction (that is, for F coinciding with &) looks like that shown in Figure 34. If F coincides with $\vee$, we obtain the table for disjunction shown in Figure 35.

In an unpublished article "A Boolian Algebra with One Constant," (1880), Peirce actually anticipated the results of M. H. Sheffer's article: "A Set of Five Independent Postulates for Boolean

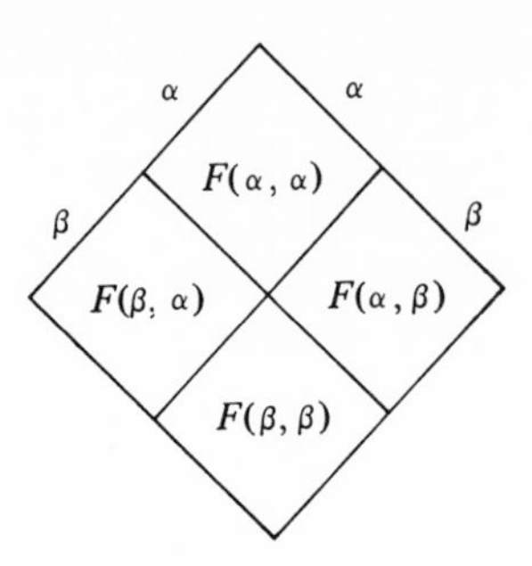

Figure 33

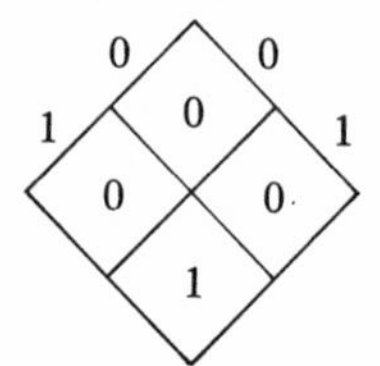

Figure 34

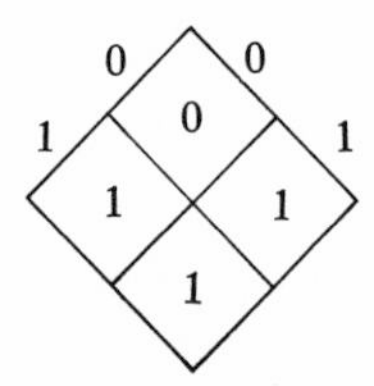

Figure 35

Algebras, with Application to Logical Constants," *Transactions of the American Mathematical Society*, vol. 4 (1913), pp. 481–488. In his unpublished article, Peirce formulates the idea of the possibility of basing a calculus of propositions on a single operation. He states that the apparatus of Boolean calculus contains the symbols 0, 1, +, −, ×, =, and > (the latter was not used by Boole but is necessary for expressing particular propositions). Peirce then suggests that it is possible to use only one symbol in place of these seven. He begins by describing the notation for conditional or "secondary" propositions ("secondary" is Boole's term): two propositions, when written as a pair, will be considered as not being fulfilled. Thus, the notation AB means that the propositions A and B are both false; the notation AA means that A is false (Reference 452, p. 13).

From this it is clear that the single constant of which Peirce spoke is nothing other than *the negation of inclusive disjunction.*

Peirce defined the negation of a proposition x in the sense of the material implication $x \rightarrow \alpha$, where α represents the identically false expression. He also gave (in 1885) a definition of the concept of tautology carried out without the traditional replacement of the moduses "use" and "reference" by the corresponding symbols (Reference 451, p. 199).

In connection with the introduction of quantifiers, Peirce thoroughly investigated the role of variables in the scientific language in general and in the formulation of mathematical theorems in particular. The calculus of predicates of the first degree was actually known to him. We also find in Peirce's work a concept related to the present-day advanced normal form (in the logic of predicates).

The logic of relations was also developed further by Peirce. He introduces the concepts of this theory in the following way. Suppose that the elements of a given finite class $\{a\}$ are denoted by the symbols $A_1, A_2, A_3, \ldots, A_{n-1}, A_n$. Then their sum is written as

$$a = A_1 + A_2 + A_3 + \cdots + A_{n-1} + A_n = \Sigma A_i,$$

where $i = 1, \ldots, n$. Suppose that a denotes "living persons," $A_1, \ldots, A_n$ are the names of persons now living, and p denotes "a group of ancestors." Then, to represent the term "a group of ancestors of some living persons," Peirce used the symbolism $\Sigma_a(p/a)$ since $p^a = p(A_1 + A_2 + A_3 + \cdots + A_{n-1} + A_n)$. For the term "the group of ancestors of every living person," Peirce used the representation $\Pi_a(p/a)$ since

$$p^a = p(A_1 \cdot A_2 \cdot \cdots \cdot A_{n-1} \cdot A_n) = pA_1 \cdot pA_2 \cdot \cdots \cdot pA_n.$$

It is easy to verify the following theorem:

$$\Pi_a\Pi_p\overline{((\Pi_a(p/a))} \vee \Sigma_a(p/a))).$$

Thus, Peirce had a very clear idea of the concept of the "composition of relations," which in contemporary notation can be represented as $aRSb$, where a and b are the names of subjects (individuals), and R and S denote relations. According to Peirce, $aRSb$ means that aRc and cSb hold for $c \in K$, where the symbol $\in$ denotes the relation of membership, and K is the symbol for the domain to which c belongs. The universal relation I is associated by Peirce with the matrix $I_{ij} = 1$ for all i, j. In Reference 442, pp. 317–378, Peirce proved, in particular, the following theorem: if $R > 0$, then $JRJ = J$.

By using a tabular definition of relations, with the aid of matrices consisting of zeros and ones, Peirce clearly anticipated the corresponding methods in Ernst Schröder's algebra of relations. Both men should be considered pioneers in the mathematical theory of structures. For example, as early as 1870 Peirce had given an axiomatic definition of a partially ordered set, defining it by using the axioms of reflexivity, nonsymmetry, and transitivity. He also formulated a definition of the sum and product in terms of the inclusion relation (Reference 442, p. 333).

In his article (Reference 443), Peirce advanced a point of view equivalent to the assertion that every structure is distributive, that is, that for every structure the relationships

$$x + y \cdot z = (x + y) \cdot (x + z) \tag{1}$$

and

$$x(y + z) = x \cdot y + x \cdot z, \tag{2}$$

hold, where $+$ denotes the operation of union, and $\cdot$ the operation of intersection of the elements of the structure. In referring to Equations 1 and 2, Peirce wrote that they are easy to prove but that carrying out the proof was too uninteresting (cited in Reference 444, p. 191, fn. 1).

However, in 1884, under the influence of Schröder's criticism, Peirce abandoned this opinion and agreed with the opposite point of view of his opponent. Nevertheless, this discussion between Peirce and Schröder was not finished. According to G. Birkhoff, Peirce later again boldly defended his initial point of view [in Reference 453 — N.S.], (cited in Reference 444, p. 191).

Peirce's logical and mathematical results were not properly evaluated in his own time, and many of them came to light only after his death. His work in semiotics was continued by C. W. Morris. Contemporary studies in semiotics are characterized by a synthesis of the Peirce tradition and the logical constructs of R. Carnap. For literature on Peirce's ideas in logic and scientific methodology, see References 427–447.

Among Peirce's disciples we should take note of the American mathematician B. I. Gilman, who concerned himself, in particular, with making Peirce's theory of simple ordering more precise (*Mind*, n.s., vol. I (1892), pp. 518–526). Thus, he formulated the condition that x cannot precede itself, though in a somewhat altered (but equivalent) form: it is impossible to have simultaneous fulfillment of the following two conditions: (1) y precedes x and (2) x precedes y. In the article "Observations on Relative Numbers with Applications to the Theory of Probability" (Boston, 1883), Gilman attempted to perfect the devices for applying Boolean algebra and ordinary algebra to the formalization of the theory of probability.

Peirce's logical ideas gradually became known on the European continent. Sometimes, however, similar ideas were proved independently. Thus, for example, many of Peirce's propositions in the logic of relations were proved again during the years 1871–1906 by the French logician J. Lachelier (1832–1918) (see References 454–456). For a characterization of the methodology of the French branch of the logic of relations, see E. K. Voyshvillo's article (Reference 457). See also References 458 and 459.

4. The Logistics of Gottlob Frege

To a large extent independently of the direction taken by Boole and Schröder, ideas of mathematical logic during the last quarter of the nineteenth century developed on the strength of the internal demands of mathematics itself (the first of these demands being the need for an axiomatic treatment of the foundations of mathematics). Here the laurels belong to the German scientist Gottlob Frege (1848–1925), who "occupied himself with the clarification of the most profound and fundamental logical relationships between the most elementary concepts and propositions of mathematics" (Reference 473, p. 6).

From 1879 to 1918, Frege was a professor at the University of Jena. A partial list of references concerning this mathematician and philosopher is presented in the bibliography (References 471–493).

Frege's name is usually associated with the emergence of a new stage in the development of mathematical logic. This stage is characterized by the axiomatic treatment of the propositional calculus, the establishment of the foundations of the theory of mathematical proof, and the formulation of the beginnings of logical semantics.

We will begin our outline of Frege's accomplishments in logic with a brief analysis of his methodological views.* Alonzo Church characterized Frege as a "consistent Platonist" (Reference 419, p. 357) or an "extreme realist" (Reference 491, p. 402); B. V. Biryukov found in Frege elements of materialism (Reference 471). In any case, we can certainly speak of Frege as a genuine antipode of the so-called "psychological" direction in logic.

In his work *The Foundations of Arithmetic. A Logico–Mathematical Study of the Number Concept* (Breslau, 1884), Frege was very critical of the views of John Stuart Mill, one of the foremost proponents of logical psychologism. In Reference 464, sec. 10, he criticized Mill's crude empirical treatment of the subject of arithmetic. In opposition to Mill's assertion that the general laws of arithmetic are inductive truths, Frege advanced the thesis that induction itself depends, for its justification, on certain general propositions of an arithmetic character. According to Frege, Mill's mistake consists in the fact that he considered the addition sign as relating to the process of physical union (Reference 469, sec. 9). Frege himself was inclined to base induction on probabilistic principles of a theoretical nature.

* P. S. Popov in Reference 368, p. 237, characterized Frege as a predecessor of Husserl and phenomenology in general. However, it is very doubtful that this is so. After all, Frege was a complete stranger to subjectivism, which is a characteristic of the phenomenological direction in philosophy. An outline of some of Frege's semantic concepts is also given by Popov (Reference 368, pp. 237–240).

In Reference 464, sec. 89, Frege criticized Kant's method of distinguishing between analytic and synthetic statements (in particular, for its unconstructive character). Frege was close to the idea that a statement should be called analytic only if it can be derived from the definitions and axioms of logic (Reference 469, sec. 3). Then a statement that is not analytic can be considered synthetic.

Thus, Frege was an opponent of both Kant's subjectivist *a priori* positions and John Stuart Mill's naive empiricism and psychologism. The divergence between Frege and Mill recalls the differences in approach to logic which can be seen from a comparison of the methodological positions of Mill and George Boole. However, unlike Boole, who through most of his work considered logic a branch of mathematics, Frege thought that the entire content of mathematics could be derived from formal logic. This idea characterizes Frege's general conception of the relationship between mathematics and logic; it came to be known in the literature as "logicism." Frege's logicism represents an attempt which is, in essence, very close to Leibniz' "universal characteristic." Frege worked in the belief that he would be able to define completely all the undefined concepts in arithmetic and prove all the unproved arithmetic axioms by reducing them to concepts and laws of logic.

The unsoundness of Frege's logicism (notwithstanding the mass of interesting logical relationships derived in his system) was demonstrated by Bertrand Russell when, at the beginning of the twentieth century, he revealed the paradoxes of the expanded calculus of predicates.*

Frege applied his logic to the problem of justifying arithmetic, which he represented as a type of formalized language. He called the logical calculus which lay at the basis of this formalized language a "calculus of concepts" (*Begriffsschrift*). By using this formalism, Frege defined, among other things, the concept of a natural number.

* Russell's discovery of the contradictions in Frege's system, of course, bore witness to the latter's failure. However, certain authors (including Russell himself) thought that logicism still could be "repaired." Only K. Gödel's famous theorem on the fundamental incompleteness of arithmetic provided a rigorous proof of the unsoundness of any form of logicism.

The basic techniques of Frege's "calculus of concepts" were presented in the book *Begriffsschrift. Eine der arithmetischen nachgebildete Formelsprache des reinen Denkens* (Halle, 1879). Here Frege dealt with the problem of graphical notation for logical concepts (ideographics). In the Introduction to Reference 460, Frege compared the relationship between his ideographics and natural language to the relationship that exists between the microscope and the eye.

Unfortunately, Frege's *Begriffsschrift* was ignored by both philosophers and mathematicians. The former were afraid of the complicated mathematical apparatus; the latter, of the use of such terminology which they, as specialists, considered "typically metaphysical" (for example, the terms "concept," "statement," "relationship," and so on). An additional difficulty arose because of the unfamiliar and complicated nature of the symbolic apparatus. The problem was that Frege's precise logical definitions were not always represented precisely enough by his devices for symbolization of logical relationships. For a long time, the extraordinary complexity of Frege's formalized language discouraged mathematicians from studying his work.* Unfortunately, only Bertrand Russell was willing to undergo the hardship of deciphering the logical works of the philosopher from Jena.†

Frege was a vigorous opponent of any definitions of concepts that contained, even to a very limited extent, references to intuition. Indicative of this tendency was his criticism of Cantor's definition of a set, according to which "by a set we will understand the union, into one aggregate, of objects which are clearly distinguished by our

* In his *Outline of the History of Mathematics*, Bourbaki said the following about Frege's language: "Unfortunately the symbolism which he [Frege] chose was insufficiently expressive, fearfully difficult to type-set, and too distant from that which was being used in mathematical practice" (Reference 531, p. 19).

† In the matter of developing a convenient method for symbolization, Guiseppe Peano and his co-workers far surpassed Frege. In 1894, they began preparing a mathematical handbook in which mathematical disciplines were to be expressed in terms of a special logical calculus. Russell was well aware of the worth of Peano's symbolism, and his own language (represented in *Principia Mathematica*, 1910–1913) was, to a large extent, a synthesis of Frege's rigorous language and the more convenient language of Peano and his school.

thought or intuition" (Reference 531, pp. 37–38). Frege correctly reproached the authors of such definitions for their vagueness and their lack of rigor and precision (Reference 465, vol. 1, p. 2).

Frege's methodological objectivism is emphasized by all contemporary historians of mathematics. For example, according to Bourbaki, Frege always "adhered to the classical position with respect to mathematical formulas, which . . . always should . . . have 'sense,'* with respect to the subconscious reality of thinking" (Reference 531, p. 49). In other words, the category of "sense" is of an objective nature for Frege; for example, a correctly constructed mathematical formula has "sense" independently of whether or not we realize that it does.

In Reference 460 Frege introduced his logical symbolism, which is of a dual nature. In the language of this symbolism, he described theorems relating to the logic of propositions as well as those relating to the logic of predicates (it should be noted that it was Frege, in 1879, who first symbolized the universal and existential quantifiers (Reference 460). He differentiated between the formulation of a statement and the judgment that this statement

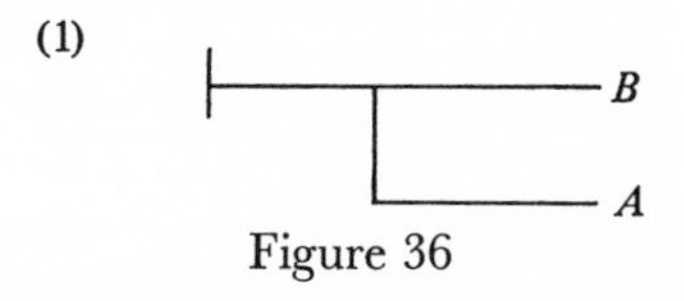

Figure 36

can be taken as proved. Thus, the symbol B denotes a statement, whereas the symbol $\vdash B$ denotes "it is judged that B holds true," where $\vdash$ is the symbol for a judgment. Frege represents the negation of the statement B as $\top B$, which corresponds to the contemporary notation $\bar{B}$ (literally, not-B).

Logical inference, treated by Frege in the sense of material implication, is represented as shown in Figure 36. Here A is the antecedent and B is the consequent, so that this symbolization is

* According to Frege, every correctly constructed ("conceived") proposition must have "sense." He considered propositions to be composite names. Frege's terminology concerning propositions is entirely of his own invention. For example, he called "the true" and "the false" "predicates of a proposition."

identical to the contemporary notation for material implication: $A \rightarrow B$. According to Frege, the apparatus of conditional statements is sufficient for formalizing causal propositions. By using the functors of implication and logical negation, Frege expressed the remaining copula of the logic of propositions. Conjunction (the

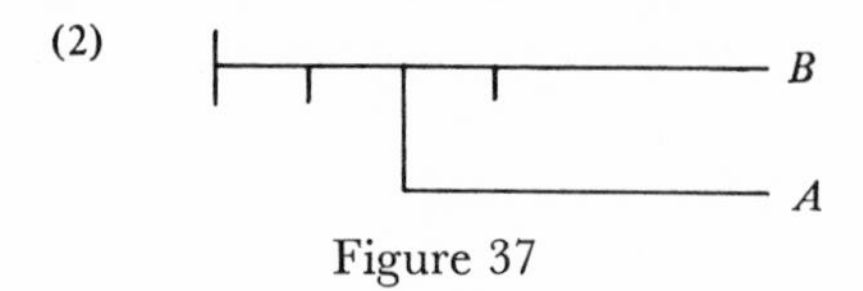

Figure 37

copula "AND") is represented by him in the form shown in Figure 37.

It is easy to see that this in fact symbolizes the expression $\overline{A \rightarrow \bar{B}}$, which is an equivalent form of representation for the conjunction A & B.

Disjunction is represented as shown in Figure 38. Here the implication $\bar{A} \rightarrow B$ is graphically represented; this is another way of

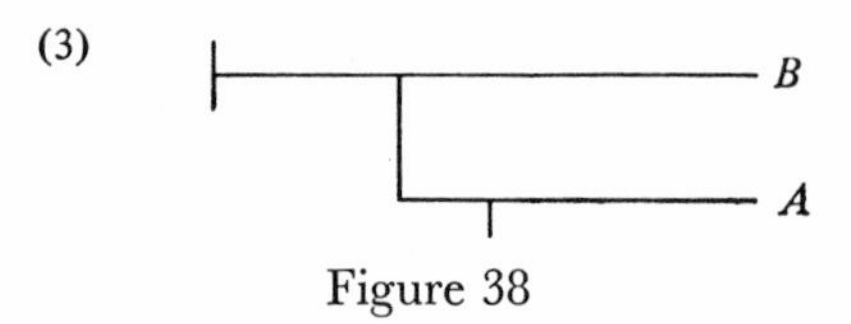

Figure 38

representing the disjunction $A \vee B$. The representation for strict (Boolean) disjunction (which, as we know, is equivalent to the conjunction of two such implications: $\bar{A} \rightarrow B$ and $A \rightarrow \bar{B}$) is somewhat more complicated, as is shown in Figure 39. The

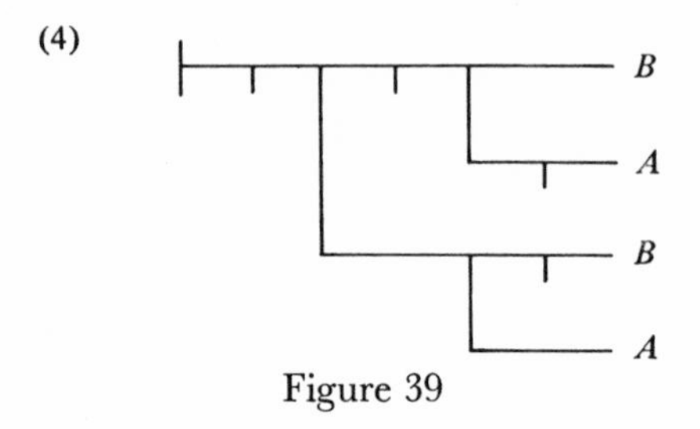

Figure 39

assertion of the identity of the names of given objects B and A was expressed by Frege in the form $\vdash(A \equiv B)$, which means "the name B and the name A have the identical conceptual content."

We will now cite Frege's symbolization for the universal quantifier. The representation

is used to denote: "for all $a, F(a)$ holds true," that is, $\forall a F(a)$ in the notation we are familiar with. As Frege himself said, "this means that the function is satisfied no matter which argument

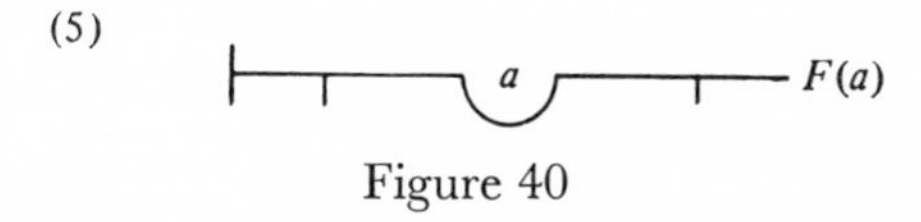

Figure 40

we choose" (Reference 460, sec. 11). The existential quantifier is expressed by use of the universal quantifier and the operation of logical negation, as shown in Figure 40. Literally: it is not true that, for all a, $F(a)$ does not hold true; that is, there exists at least one a which has the property F. For the way Frege displayed the well-known syllogism "from the universal to the particular" $\forall a F(a) \rightarrow F(b)$, see Figure 41. And Figure 42 shows how he

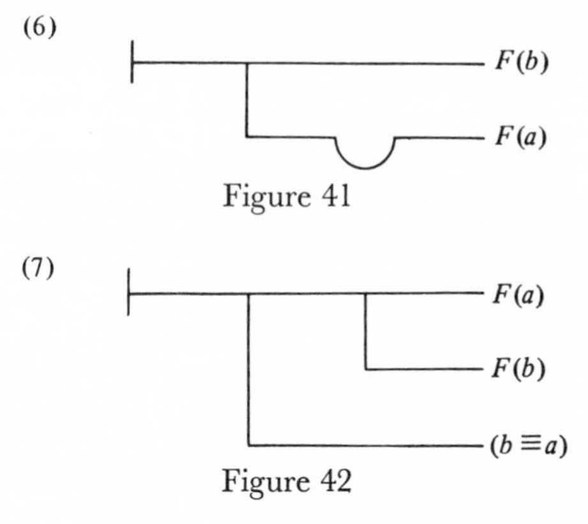

Figure 41

Figure 42

symbolized the Leibnizian definition of equality, which is equivalent to the contemporary expression $(b \equiv a) \rightarrow (F(b) \rightarrow F(a))$. This definition was given by Frege in Reference 464 in 1884. In 1898, Reference 465, the principle of identity was defined by Frege by use of quantifiers of the predicate, a definition which corresponds to the implication

$$(\varphi(x \equiv y)) \rightarrow (\varphi(\forall F(F(x) \rightarrow F(y)))). \tag{8}$$

Already in Reference 460 Frege had excluded from use the unclear concept of a "variable quantity" and had replaced it with a precisely defined concept of a variable as a symbol of a fixed type. In generalizing the traditional concept of a function, he developed the thesis (related to Boole's views) that the values of a function should not be strictly limited to numbers and that there even should exist functions whose range of definition is arbitrary.

In Frege's logical method, the symbols lying at the foundation of a formal system are subdivided into (1) initial logical terms and (2) undefined (in terms of mathematical concepts) terms of arithmetic. Analogously, he distinguished logical axioms (which do not include any other terms except logical ones) and postulates (which include arithmetic terms as well as logical terms). Frege constructed the first axiomatic system for the propositional calculus, based only on implication and negation as the undefined logical functors (Reference 460). This axiomatization was of the following form (for simplicity we present it in contemporary notation):

$$A \rightarrow (B \rightarrow A), \tag{1}$$

$$(C \rightarrow (A \rightarrow B)) \rightarrow ((C \rightarrow A) \rightarrow (C \rightarrow B)), \tag{2}$$

$$(A \rightarrow (B \rightarrow C)) \rightarrow (B \rightarrow (A \rightarrow C)), \tag{3}$$

$$(A \rightarrow B) \rightarrow (\bar{B} \rightarrow \bar{A}), \tag{4}$$

$$\bar{\bar{A}} \rightarrow A, \tag{5}$$

$$A \rightarrow \bar{\bar{A}}. \tag{6}$$

The first axiom expresses the law of assertion of the consequent of an implication; the second expresses the law of self-distributivity of the functor $\rightarrow$; the third formulates the law of commutativity; the fourth is the principle of contraposition; the last two axioms (the fifth and sixth) taken together express the theorem that states the equivalence of assertion and double negation. From the system of Axioms 1–6,* Frege formally derived many of the theorems of propositional calculus, including, for example, the following:

$$A \rightarrow (B \rightarrow \overline{(A \rightarrow \bar{B})}), \tag{7}$$

$$A \rightarrow (\bar{A} \rightarrow B). \tag{8}$$

The laws of deduction used by Frege were the substitution of equivalents for equivalents and *modus ponens*. Unfortunately, Frege's axiomatic treatment of the propositional logic was not understood until long after its publication in his *Begriffsschrift* (1879). Thus, the specific relationships of the propositional calculus developed within the framework of the original approach taken by Boole and his followers.

In introducing quantifiers and a concept closely related to that of a propositional function (*Wertverlauf einer Funktion*), Frege accomplished the transition from a purely propositional calculus to a logic of predicates. On the basis of this foundation, he intended to formalize the entire content of arithmetic. Naturally, he had to begin with an attempt at a logical definition of a quantitative number. Of course, he attempted to define the concept of a number in such a way that its definition would be independent of any concepts other than those already related to the interpretation of

* This system of axioms satisfies the consistency criterion. However, as J. Lukasiewicz proved (Reference 488), it is not independent, for Axiom 3 logically follows from the conjunction of the first two axioms. In connection with this, Lukasiewicz in Reference 488 proposes his own variant of an axiomatization for the propositional calculus (using implication and negation). Lukasiewicz's axiomatization is a simplification of Frege's and consists of the following three axioms:

$$(A \rightarrow B) \rightarrow ((B \rightarrow C) - (A \rightarrow C)) \tag{I}$$

$$A \rightarrow (\bar{A} \rightarrow B) \tag{II}$$

$$(\bar{A} \rightarrow A) \rightarrow A. \tag{III}$$

his calculus as a logical system. At the core of Frege's definition of a number lies his use of the concept of a one-to-one correspondence (*beiderseits eindeutig*). In his *Grundlagen der Arithmetik*, we read:

"1. The expression: 'The concept F is numerically equivalent (*gleichzahlig*) to the concept G' denotes the same thing as the expression 'there exists a relation f which establishes a one-to-one correspondence between the objects falling under the concept F, on the one hand, and the objects falling under the concept G, on the other.' 2. The number which belongs to the concept F is the extent of a concept which is 'numerically equivalent to the concept F.' 3. 'N is a number' denotes the same thing as 'there exists a concept such that N is the number which belongs to that concept.'" (Reference 464, sec. 72.)

It is interesting to note that the expression "numerically equivalent" used in this definition is identical in meaning to the concept of "equipotence of sets" (set equivalence) used by the founder of set theory, Georg Cantor.

Thus, a "natural number" was considered by Frege a "property of a concept" (Reference 464, sec. 57) and not a property of a subject. In other words, a number is a certain predicate of a predicate. Frege specifically emphasized that the concept of "one-to-oneness" of a correspondence does not at all imply the use of the concept "one," and hence the argument is not a vicious circle. Incidentally, Frege also established a rigorous differentiation between a subject and a class which consists of this single subject.

In Reference 464, sec. 68, we find Frege's definition of the cardinal number of a set A as the set of all sets equivalent (equipotent) to that set A. However, it is this fundamental definition that proves to be the "Achilles' heel" of Frege's formalism: the problem is that the set of sets equivalent to the set A possesses paradoxical properties. This latter fact was established in 1905 by Bertrand Russell, when he revealed the contradiction inherent in the concept of the set of all sets which do not contain themselves as elements.

The logical paradoxes in Frege's system turned out to be the stumbling block in his attempt at a justification of mathematics. He tried to prove the consistency of arithmetic by constructing for it models which, in essence, lead directly to the problem of the consistency of Cantor's theory of sets. Later, even David Hilbert avoided any attempts to prove the consistency of arithmetic by the method of models.

In his theory of proper names in the broad sense, developed in particular in *Über Sinn und Bedeutung* (Reference 463), Frege showed himself as a pioneer in semantic studies.* The title of this well-known work (published in 1892) already gave some indication of its importance. Frege made a sharp distinction between the concepts sense (*Sinn*) and meaning (*Bedeutung*), as applied to the names of subjects. In his opinion, the "meaning" of a name is the subject (*Nominatum*) called by that name. Frege's use of the term *Nominatum* for the object of a naming relation is identical to Church's use of the term "denotant" (Reference 491, par. 01). As for the "sense" of a name, this can be defined as information about the subject (*Nominatum*) of the name. By assimilating this information included in the name, a person penetrates into its sense.

Names can have one and the same Nominatum but different senses. For example, the names "sin 150°" and "$\frac{1}{2}$" have the exact same Nominatum (that is, they name one and the same subject — a specific real fractional number), but the sense of these names is different ("sin 150°" contains trigonometric information, whereas "$\frac{1}{2}$" conveys information relating to the realm of arithmetic). Frege presented the following example (Reference 463). Suppose we draw the equilateral triangle *ABC* with medians *AD*, *EC*, and *BF* intersecting at the point *Q* (see Figure 43).

Let us now consider the sense and meaning of the following two names: (1) "the point of intersection of *AD* and *EC*"; (2) "the point of intersection of *EC* and *BF*." It is clear that expressions 1 and 2 determine the same point *Q* (that is, their "meaning" is the

* Already in *Über Funktion und Begriff* (1891), Frege treated "true" and "false" as abstract subjects. According to this approach, all true statements have the same subject — "the true" (*das Wahre*), and all false statements also have the same subject — "the false" (*das Falsche*). A function with a range of values consisting only of "the true" and "the false" is, according to Frege, a function of a statement (that is, a "propositional function" in Russell's logic). Frege treats a function of a statement as the analog of a numerical function in mathematical analysis (Reference 491, p. 359).

same); however, their sense is not identical since the methods of definition of the point Q are different.

Another of Frege's examples: The names "Evening star" and "Morning star" have different senses but the same meaning (since both names refer to the same Nominatum — the planet Venus). Frege takes special note of the class of names which have

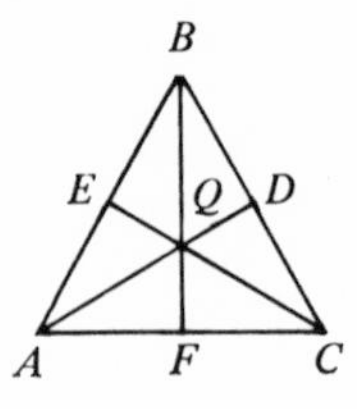

Figure 43

sense, but are devoid of meaning (because of the absence of a Nominatum); this is the case, for example, with a name like "Pegasus."

One of Frege's special concerns was the clarification of the question of applicability of the so-called thesis of extent (volume) to statements which include proper names. In this connection, he distinguishes between "direct" and "indirect" use of a name. This is evident from the following text from Frege: "In the case of direct (*gewöhnlich*) use of names, the subject of the naming (Nominatum) is the meaning of the name. However, we may have situations in which the intention is to speak about the names themselves or about their senses. This happens, for example, when we quote the words of another person . . . In this case, the proper names denote primarily the words of that other person, so that only these words are used directly. Here we are dealing with signs of signs. Thus, in the written language, the corresponding notations for names are included in quotation marks. Because of this graphical inscription enclosing the names, it is inadmissible to consider them from the point of view of their direct use" (Reference 461, p. 28).

For example, in the statement (1) "Tegucigalpa is the capital of Honduras," both the subject (Tegucigalpa) and the predicate (the capital of Honduras) are used directly. However, in the statement

(2) "Student Ivanov wondered: 'Is Tegucigalpa the capital of Honduras?'", the same names (Tegucigalpa and capital of Honduras) are used indirectly (*ungerade* in Frege's terminology). Since Tegucigalpa is in fact the name of the capital of Honduras, we can consider the corresponding names to be mutually interchangeable (that is, we can apply to them the thesis of extent, for concepts, since they have the same Nominatum). After such a transformation, Statement 1 becomes a tautology: (1*) "Tegucigalpa is Tegucigalpa." While Phrase 1 contains geographical information, Phrase 1*, obtained from Phrase 1 by application of the thesis of extent, obviously does not contain any useful information. However, both Phrase 1 and Phrase 1* are true phrases. Let us consider what happens if an analogous interchange of equal extent for equal extent is carried out on Statement 2. As a result, we would obtain the phrase: (2*) "Student Ivanov wondered: 'Is Tegucigalpa Tegucigalpa?'" Phrase 2* would naturally be considered false. The reason for the fact that here the thesis of extent leads to a fiasco must be sought in the specific character of the indirect use of proper names.† This specific nature results in the following circumstance: the Nominatum in indirect usage of the name becomes the "sense" of the name in direct usage.

The difficulties associated with the naming relation are still a subject of lively discussion in contemporary logical semantics, and Frege's ideas about semantics are still rather timely.‡

† Notwithstanding, Frege did not reject the thesis of extent but, instead, tried to make it more precise, giving the following formulation: "Under the conditions of validity of our supposition that the 'meaning' of a *proposition* is its truth value, this truth value is preserved when one part of the proposition is replaced by an equivalent expression with the same meaning but with a different sense" (Reference 463, p. 35). This point of view can be considered a generalization of the rule of substitution of equals, formulated as far back as Leibniz (and also a refinement of the substitution principle in Jevons' treatment). Incidentally, we should note that in Frege's view the Nominatum of a proposition is in its truth value, whereas the "sense" of the proposition reduces to the judgment contained in it. The proposition considered as a whole is the name of a truth value.

‡ Many of the aspects of Frege's semantics which are acceptable in themselves were generalized in Reference 491 by Alonzo Church. Church considered that the weak point of Frege's conception was that it did not include any theory concerning the equality of "senses."

5. Giuseppe Peano and His School

The ideas of the well-known Italian mathematician Giuseppe Peano (1858–1932) were very close to the line of thought originated by Gottlob Frege. As we know, one of the most important problems in the method of axiomatic systematization of mathematics consists in exhibiting all the admissible methods of deriving new propositions from the propositions of the given axiomatization. Peano first turned to precisely the solution of this problem working independently of Frege but essentially within the same framework. It is true that we do not find in Peano's work a logical method in finished form (Reference 532, p. 149). Rather, the significance of Peano's total achievement lies in its being a transitional link between the algebra of logic (in the form given by Boole, Schröder, Peirce, and Poretskiy) and the contemporary form of mathematical logic.

Peano's results had definite national roots. Let us briefly discuss those Italian scientists who, prior to Peano, were, to one degree or another, close to logico-mathematical conceptions (see References 495–501).

1. Girolamo Saccheri (1667–1733) displayed great interest in the development of the techniques of logical proof (see his *Logica demonstrativa*, Turin, 1697, which was published anonymously. The third edition of the book, Cologne, 1735, revealed the author's name). However, his most interesting work is *Euclides ab omni naevo vindicatus* (Milan, 1733). This book allows us to see Saccheri as a distant forerunner of ideas which are characteristic of conceptions of non-Euclidean geometries. Saccheri began with an attempt to prove Euclid's fifth postulate on parallels. He began by considering a quadrangle whose opposite sides are equal and whose base angles are each equal to d. He first proved that the upper angles must be equal to each other and then successively analyzed the following three suppositions: (1) each of these angles is greater than d; (2) each of these angles is less than d; (3) each of these angles is equal to d.

From Hypotheses 1 and 2 he derived contradictions, which led him to take Supposition 3 as proved. As G. Vileytner noted in Reference 286, p. 361, the subtlety of Saccheri's arguments and the keenness of his logical calculations are a result of his noticing that these hypotheses are equivalent to the following possibilities for the angles of a triangle:

$$\Sigma > 2d; \quad (1)$$

$$\Sigma < 2d; \quad (2)$$

$$\Sigma = 2d. \quad (3)$$

2. Ludovico Richeri (eighteenth century), in an article (Reference 497) in 1761 showed himself to be under the influence of Leibniz' ideas about a "universal characteristic" and applied pasigraphy. This article attracted the attention of J. H. Lambert in *Nov. Ac. Erud.* (1767), pp. 334–344.

3. The Neapolitan F. Pagano turned to the development and dissemination of ideas concerning the logic of probability (Reference 498).

4. A persistent attempt to reform traditional logic was made by Pasquale Galuppi (1770–1846). (See, for example, References 498 and 500.) From 1831 on, he taught logic in Naples. In the realm of methodology, he relied on Leibniz, criticized Kant, and attempted a critical reworking of the results of Scottism and French Sensualism (Reference 501).

Let us now turn to a characterization of the logical accomplishments of Peano and his school. Peano's fundamental results were published in the five-volume *Formulaire de Mathématiques* (Turin, 1895–1905). From among the works of Peano that are of interest from the point of view of mathematical logic, we cite References 502–508. For literature on Peano and his school (works of his followers are cited in References 509–518), see References 519–535.

We will begin by noting Peano's contribution to the development of logical symbolism and terminology. In particular, he introduced the following symbols (which, of course, are still used today):

1. $\in$ (read "is included in"; used to express the relation of membership of an element in a set).

2. $\supset$ (used to express inclusion of one set in another).

3. $\cup$ (used to symbolize non-Boolean "inclusive," union of sets).

4. $\cap$ (used to denote the operation of set intersection).

The last two signs are also widely used in the theory of structures (see, for example, Reference 331).

Peano also introduced the now widely used (although in adapted forms by Whitehead, Russell, and others) system of symbolic notation for so-called sentential copulas (which serve for the formation of new propositions from a number of given ones, which, in a particular case, may even be equal to only one). Among those of Peano's innovations that did not gain acceptance, we note, in particular, the symbol $\imath$ (a lower-case italic i without the dot) as denoting a quantifier with respect to the predicate. Peano applied this to expressions having one- or two-sided implicative form (that is, of the form $A \rightarrow B$ or of the form $A \rightleftarrows B$). The first scientist to modify Peano's symbolism for quantifiers was Bertrand Russell.

Being extremely pedantic with respect to matters of notation, Peano began to apply a method of systematic enumeration of logico–mathematical propositions: axioms, lemmas, theorems, etc. (here he was inspired by the system of indices used for decimal classification of documents). He also began the systematic use of dots as abbreviations for numerous parentheses. For example, instead of $A \rightarrow (B \rightarrow A)$, the Italian mathematician would write $A \Rightarrow . B \Rightarrow A$, where the symbol $\Rightarrow$ expresses the relation of implication of one proposition by another.

In 1897, Peano drew a distinction between an "actual" and an "apparent" variable. His terminological distinction is preserved in a corresponding pair of contemporary terms: "free" and "bound" variable (see, for example, Reference 360, pp. 85–86); for some semantic inconveniences related to the use of the term "actual variable" in mathematical practice, see Reference 532, p. 349. In 1887, Peano began to use the term "measurable set" (in contemporary terminology, "integrable set"); he also began to express the concepts of elementary geometry in set-theoretic language.

Peano's most important contribution to the development of the theory and practice of the axiomatic method was his system of axioms for the arithmetic of the natural numbers (1889, 1895). As the undefined concepts, he used (1) "the set of natural numbers," (2) "zero," (3) "subsequent" (that is, the predicate "directly follows after the given number," Reference 503). The substance of Peano's axiomatization can be formulated as follows.

Axiom 1. No number has zero as a subsequent.

Axiom 2. Two different numbers have two different subsequents.

Axiom 3. If some property T holds for zero and if, whenever T holds for any number x, it also holds for the subsequent of x, then T holds for every number n.

Let $\{N\}$ denote the set of natural numbers and $S(x)$ the operation of generating the number directly following the given one (that is, $S(x) = x + 1$). Then Peano's axiomatization (Reference 507, vol. 2, sec. 2) can be written in present-day logical notation as follows (0 is the symbol for arithmetic zero):

1. $\forall x((x \in \{N\})S(x) \neq 0)$,
2. $\forall x \forall y((x \in \{N\} \;\&\; y \in \{N\})((x \neq y) \rightarrow (S(x) \neq S(y))))$,
3. $((x \in \{N\}) \;\&\; (y \in \{N\})(T(0) \;\&\; \forall x \forall y((T(x) \;\&\; S(x, y)) \rightarrow T(y)))) \rightarrow \forall x T(x)$.

It is clear that Peano actually proceeded from the inductive definition of the binary predicate for addition (which we will denote by S^*). This definition of S^* consists of the equations

$$S^*(y, 0) = y \tag{1}$$

and

$$S^*(y, Sz) = S(S^*(y, z)), \tag{2}$$

where S, as previously, denotes "has as a subsequent." Later, inductive definitions found very wide application in structural and mathematical linguistics.

On the basis of his axiomatization, Peano constructed the entire theory of natural numbers. In particular, he showed how the elementary theorems of arithmetic can be obtained from his axioms (Reference 507). It should be noted that Peano's axiomatization was to a significant degree inspired by the ideas expressed by R. Dedekind in his treatise (Reference 534). In general, Peano was under the strong influence of the theory of functions developed by Dedekind (Reference 531, p. 20).

Peano's axiomatization is characterized by the fact that it is categorical, in that it possesses completeness of content. In other words, in substantive arithmetic Peano's axioms determine the system of natural numbers, up to an isomorphism. The question of proving the independence of the axioms is taken up in Reference 508. Peano presented a method for such a proof. The method proceeds as follows. In order to show that a given axiom of the axiomatic system being considered is not a logical consequence of the remaining axioms of the same system, Peano constructs a model in which all the axioms are fulfilled, except the one whose independence is to be proved.

The system $\langle\{N\}, 0, S\rangle$ later came to be called a Peano model, and it was proved that every Peano model is an induction model but that the converse is not true (see Reference 530, p. 23, fn. 2). The first critic of the Peano model was H. Poincaré, who was convinced of the nonformalizability of the principle of complete mathematical induction (see Peano's third axiom). Unfortunately, Poincaré's authority shook the confidence of mathematicians in Peano's model, and its rehabilitation did not come quickly.

It is completely understandable that Peano stopped considering arithmetic the foundation of the discipline of mathematics. Instead he turned to a new science, related to set theory, upon which he hoped to erect the remaining structure of mathematics.

Peano also concerned himself with mathematical ideographs (Reference 535) and problems of Esperanto based on Latin (Reference 526); he actually took part in the development of the formal logical theory of definition by means of abstraction. Feys considers Peano one of the forerunners of combinatorial logic (Reference 523).

Peano worked in close cooperation with the Italian logicians and mathematicians Giovanni Vailati, Giovanni Vacca, C. Burali-Forti, and Alessandro Padoa. In part, Peano determined the direction of the investigations of these men; in this sense, we can speak about Peano's school. Let us briefly characterize the work of these four of its representatives:

1. Giovanni Vailati (1863–1909) had a very wide circle of interests. In Reference 515 he reviewed L. Couturat's study on

Leibniz' logic. In Reference 514, he proposed a definition of a simple ordering in terms of the relation "precedes" by using a system of three postulates. Vailati also defined the concept "between," which is undefined in Peano's work; he thus established one of the possible forms of axiomatization for ordering the elements of a so-called closed series (the points of a circle constitute such a closed series).

An object of constant attention for Vailati was the propositional law $(\bar{a} \rightarrow a) \rightarrow a$; he investigated in great detail the history of the discovery and application of this law.

In Reference 516, Vailati established the following propositional theorem: If $(a \,\&\, b) \rightarrow (c \,\&\, d)$ and $(b \vee c) \rightarrow (a \vee d)$, then $b \rightarrow d$. This can be called Vailati's theorem. To prove Vailati's theorem, we use the ordinary devices and algorithms for elimination in the Poretskian sense (Vailati himself proved the theorem by another method). Reducing the premises $(a \,\&\, b) \rightarrow (c \,\&\, d)$ and $(b \vee c) \rightarrow (a \vee d)$ to disjunctive normal form, we obtain

$$(\bar{a} \vee \bar{b} \vee (c \,\&\, d)) \,\&\, ((\bar{b} \,\&\, \bar{c}) \vee a \vee d),$$

from which, by removing brackets and simplifying, we obtain

$$(\bar{b} \,\&\, \bar{c}) \vee (a \,\&\, \bar{b}) \vee (\bar{a} \,\&\, d) \vee (\bar{b} \,\&\, d) \vee (c \,\&\, d).$$

Now we replace the letters to be eliminated (a and c) with the identically true expression T (in accordance with Poretskiy's algorithm). Then we have

$$(\bar{b} \,\&\, \mathrm{T}) \vee (\mathrm{T} \,\&\, \bar{b}) \vee (\mathrm{T} \,\&\, d) \vee (\bar{b} \,\&\, d) \vee (\mathrm{T} \,\&\, d),$$

from which we obtain $\bar{b} \vee d$, that is, $b \rightarrow d$, which exactly coincides with Vailati's result.

2. Giovanni Vacca was interested in problems of the history of logic. As examples, we can cite his article (Reference 512) on the logic of Leibniz, or his evaluation of the mathematical logic of J. D. Gergonne in Reference 513.

3. C. Burali-Forti is known primarily as the author of the paradox labeled with his name. In an article (Reference 509) in 1897, he denied the legitimacy of assuming the existence of a set of ordinal numbers. After all, asserts Burali-Forti, such a set would

appear to be well ordered, so that it can be mapped one to one onto its proper subset; but this leads to an absurdity. Therefore, any satisfactory theory of ordinal numbers must be constructed in such a way that Burali-Forti's paradox does not occur in it. As an example of a theory which eliminates this paradox, we can cite, for instance, that presented in Reference 533.

Burali-Forti was also the author of a handbook on logic (Reference 510), published in Milan in 1894 but almost forgotten today.

4. Alessandro Padoa occupied himself with a detailed study of the question of determinateness. He proposed the original method for solving the problem of the dependence or independence of a system of initial undetermined concepts. Padoa can be considered a forerunner of the theory of recursive definitions (the latter are constructed like the definition of the predicate of addition in Peano's model); see Padoa's articles, References 517 and 518.

Following Dedekind, Padoa worked out a theory of mappings. In particular, it was he who introduced the example of the mapping $x \leftrightarrow x + 2$.

Conclusions

The material we have presented allows us to draw a number of conclusions, of both a general and a specific nature.

1. The process of estalishment of the ideas of mathematical logic cannot be likened to a continuously rising curve. Periods of progress in the development of logic sometimes alternated with periods of relative regression. This uneven progress, which is not unexpected in principle, once again reaffirms the Marxist–Leninist thesis that the development of science takes place not evenly but by means of a process of forward leaps, whose roots lie in qualitative revolutions in the content and methods of science.

2. The formulation of the individual positions taken by mathematical logic in medieval Europe occurred, in the main, in the work of scientists who occupied nominalist positions, which, to a greater or lesser extent, were close to materialism. The principal results of these scientists related to the theory of logical inference and to semantic antinomies.

3. The difficulties which arose in the later investigations of G. W. Leibniz are often explained by reference to the intensional nature of his logic (L. Couturat) or to his tendency to express in his calculus all the positions of the Aristotelian formalism (N. Rescher). However, this author believes that Leibniz' difficulties were basically a result of the absence, in his logic, of relationships that directly linked direct and inverse logical operations (for example, logical multiplication and logical division).

4. The fundamental traits which characterize the line of development from Leibniz to George Boole consist of the following: the treatment of logical rules determining logical operations as only *analogs* of corresponding laws for mathematical operations; the accompanying lack of verification in extrapolating mathematical

methods to formal logic, and the overestimation on the part of the pioneers of mathematical logic of the analogies existing between logical transformations and mathematical operations.

5. George Boole considered his calculus an auxiliary instrument for the solution of problems in the theory of probability. His logical system (which must be distinguished from contemporary Boolean algebra) cannot be directly classified as a ring with idempotent elements. However, in our work we have demonstrated the possibility of insertion of Boole's calculus in a ring of residues modulo two since the operation of logical division can be eliminated by using the formula

$$\frac{X}{Y}(v) \equiv X \oplus v(1 \oplus Y),$$

where $X \leqslant Y$ and where $\oplus$ is the symbol for symmetric difference and v is an indeterminate.

It is also necessary to introduce a number of corrections to Boole's rules of interpretation for the expanded expression of a given system of premises.

6. In addition to its applications in the algebra of logic, the language of minimal normal forms developed by P. S. Poretskiy can be effectively applied for the simplification and design of structural formulas for weak-current relay-contact networks.

7. The logical and algebraic studies of the originator of semiotics, C. S. Peirce, are founded on the methodology of his *pragmaticism*, which many authors usually consider to be the same as pragmatism. However, Peirce's pragmaticism is radically different from the doctrine of the "pragmatists" and should be considered as a movement which is closer to materialism.

In certain areas, Peirce's algebra of relations is more subtle than the corresponding theory of E. Schröder. However, in general, it falls short of Schröder's theory in systematization and extent of development.

References

Medieval logic

1. Boehner, P. "Ockham's Theory of Supposition and the notion of truth." *Franciscan Studies*, 1946, pp. 261–292.
2. ———. *William Occam*. Manchester, 1945.
3. ———. "Investigations in the Logic of William Occam." *Franciscan Institute Publications*, sec. 3 (1945), pp. 58–88.
4. Salamucha, J. "William Ockham's logic of propositions." *Franziskanische Studien*, vol. 32 (1950).
5. Scholz, H. "Review of E. A. Moody's 'The Logic of William of Ockham.'" *Deutsche Literaturzeitung*, vol. 58 (1937).
6. Thornay, Stephen Chak. *Ockham, Studies and Selections*. Illinois, 1938.
7. ———. *The Nominalism of William Ockham*. Chicago, 1936.
8. Scholz, H. "Review of Boehner's article on Ockham's logic." *Deutsche Literaturzeitung*, vol. 69 (1948).
9. Bainberg, U. P. "Review of Boehner's article on Ockham's 'Treatise on Predestination.'" *The Philosophical Review*, vol. 56 (1947).
10. Hochstetter, Erich. *Studien zur Metaphysik und Erkenntnistheorie Wilhelms von Ockam*. Leipzig, 1927.
11. ———. *Historisch-philosophisches Handbuch zur Geschichte der Philosophie* (Leipzig, 1879), S.V. "Nicolettus."
12. Popov, P. S. *Logika Aristotelya i logika formal'naya* [The logic of Aristotle and formal logic]. Series in History and Philosophy, vol. 2, 1945.
13. Styazhkin, N. I. "Elementy algebry logiki i teorii semanticheskikh antinomiy v pozdney srednevekovoy logike "[Elements of the algebra of logic and the theory of semantic antinomies in later medieval logic]." In *Logicheskie issledovaniya* [Studies in logic]. Moscow, 1959.
14. Bezobrazova, M. V. "O 'Velikoy Nauke' Raymunda Lulliya v russkikh rukopisakh XVII vv." [On the "Great Art" of Raymond Lully in Russian seventeenth-century manuscripts]. *Zhurnal*

Ministerstva Narodnogo Prosvesheheniya, vol. II, February 1896, pp. 383–389.
15. Sokolov, N. A. "Filosofiya Raymunda Lulliya i ego avtor" [Raymond Lully and his philosophy]. *Zhurnal Ministerstva Narodnogo Prosvesheheniya*, August 1907, pp. 331–337.
16. Akhmanov, A. S. "Logichskie ucheniya v devney Gretsii i drevnem Rime" [Studies in the logic of ancient Greece and ancient Rome]. Dissertation, part 1, Moscow State University, 1950. Mimeographed.
17. Prantl, Karl. *Geschichte der Logik im Abendlande*, vols. I–IV. Leipzig, 1855–1870.
18. Vladislavlev, M. I. "Skholasticheskaya logika" [Scholastic logic]. *Zhurnal Ministerstva Narodnogo Prosvesheheniya*, part XII, sec. 2, no. 3, 1872, pp. 195–271.
19. ———. "Istorischeskiy ocherk logiki Aristotelya, skholasticheskoy, formal'noy i induktivnoy" [Historical outline of Aristotelian, Scholastic, formal and inductive logic]. Foreword to *Logika* [Logic]. St. Petersburg, 1881.
20. Moody, E. A. *Truth and Consequence in Medieval Logic*. Amsterdam, 1953.
21. Boehner, P. *Medieval Logic: An Outline of Its Development from 1250 to 1400*. Manchester, 1952.
22. Akhmanov, A. S. "Formy mysli i zakony formal'noy logiki" [Forms of thought and laws of formal logic]. In *Voprosy logiki* [Problems in logic]. 1955.
23. Weyl, G. *O filosofii matematiki* [On the philosophy of mathematics]. Moscow, 1934.
24. Serrius, S. *Opyt issledovaniya znacheniya logiki* [Experiment in the investigation of the meaning of logic]. Moscow, 1948.
25. Prantl, K. M. *Psellus und P. Hispanus*. Leipzig, 1867.
26. Boehner, P. "Bemerkungen zur Geschichte der Morganschen Gesetze in der Scholastic." *Archiv für philosophie*, September 1951.
27. Lukasiewicz, J. *Aristotle's Syllogistic from the Standpoint of Modern Formal Logic*. Oxford, 1951.
28. Lakhuti, D. G. "Sistema strogoy implikatsii V. Akkermana i problema formalizatsii logicheskogo sledovaniya" [The system of strict implication of W. Ackermann and the problem of formalization of logical implication]. Dissertation, Moscow State University, 1957. Mimeographed.
29. Maier, A. "Zwei Grundprobleme der scholastischen Naturphilosophic." 2nd ed. *Storia e letteratura*, vol. 37. Rome, 1951.
30. Nelson, N. *Peter Ramus and the Confusion of Logic, Rhetoric and Poetic*. Michigan, 1947.
31. Dürr, Karl. "Logika predlozheniy v Sredniye Veka" [The

logic of propositions in the Middle Ages]. *Erkenntnis*, vol. 7, no. 3 (1938).
32. Bendiek, Johannes. "Skholasticheskaya i matematicheskaya logika" [Scholastic and mathematical logic]. *Franziskanische Studien*, vol. 31 (1949).
33. Birkenmaier, Alexander. "Retsenziya na rabotu Ya. Salamukhi: Vozniknoveniye problemy antinomiy v srednevekovoy logike" [Review of Salamucha's article: 'The emergence of problems of antinomies in medieval logic']. *Kwartalnik filozoficzny*, vol. 15 (1938).
34. Schneid, A. *Aristoteles in der Scholastik*. Berlin, 1875.
35. Bochenski, I. M. "de consequentiis scholasticorum earumque origine." *Angelicum*, vol. 15, (1938), 92–109.
36. Salamucha, Jan. "Review of Bochenski's book 'On the history of the logic of modal relations.'" *Kwartalnie filozoficzny*, vol. 15 (1938). (Polish).
37. Mordukhaj-Boltovskoj, Dmitrij. *Insolubiles in scholastica et paradoxos de infinito de nostro tempore*. Warsaw, 1939.
38. Grabmann, Martin. *Bearbeitungen und Auslegung der aristotelischen Logik aus der Zeit von Peter Abaelard bis Petrus Hispanus Mitteilungen aus Handschriften deutscher Bibliotheken*. Berlin, 1937.
39. Bernal, J. *Nauka v istorii obshchestra*. [Science in the history of society] (Russian translation). Chap. 6, Moscow, 1956.
40. *B.S.E.*, s. v. "Skholastika" [Scholastics].
41. Tseytin, Z. A. "Skholasticheskiy empirizm" [Scholastic empiricism]. *P.Z.M.*, nos. 8, 9, 1924.
42. Grabmann, M. *Die Geschichte der scholastischen Methode*, vols. 1, 2. Berlin, 1956.
43. Trakhtenberg, O. V. *Ocherki po istorii zapadno-evrepeyskoy filosofii* [Outline of the history of Western European philosophy]. Moscow, 1957.
44. Boehner, P. "Srednevekovoy krizis logiki" [The medieval crisis of logic]. *Franciscan Studies*, vol. 25, (n.s. vol. 4, 1944).
45. Bochenski, I. M. "Ob analogii" [On analogy]. *The Thomist*, vol. 11 (1948), pp. 424–447.
46. Dürr, Karl. "Alte und neue Logik." *Jahrbuch der schweizerischen Philosophischen Gesellschaft*, vol. 2 (1942), pp. 104–122.
47. Boehner, P. "The system of Scholastic logic." (Spanish translation by M. R. Krespo) *Revista de la Universidad Nacional de Córdoba* (Argentina), vol. 31 (1944), pp. 1599–1620.
48. Roos, H. "Martinus de Dacia und seine Schrift: De modis significandi." *Classica et Medievalia*, VIII, 1 (1946).
49. Wilson, C. *Medieval Logic and the Rise of Mathematical Physics*. Madison, Wisconsin, 1956.

50. Kneale, W. and M. Kneale. *The Development of Logic*. Oxford, 1962, chap. 4.
51. Bochenski, I. M. *Ancient Formal Logic*. Amsterdam, 1953.
52. Biryukov, B. V. "Kak voznikla i razvivalas' matematicheskaya logika" [The origins and development of mathematical logic]. *Voprosy Filosofii*, no. 7 (1959).
53. Bochenski, I. M. *A History of Formal Logic*. Notre Dame, Indiana, 1961.
54. Zinov'ev, A. A. "Rasshiryat' tematiku logichekikh issledovaniy" [To widen the thematics of logical studies]. *Voprosy Filosofii*, no. 3 (1957).
55. Yakubanis, G. "Otsenka sillogizma v glavneyshie momenty ego istorii" [Evaluation of the syllogism at the principal moment in its history]. *Kievskie universitetskie izvestiya*, no. 9, 1910.
56. Werner, K. *Die Sprachlogik des Duns Scottus*. Vienna, 1877.
57. Gyunter, Z. *Istoriya estestvoznaniya v drevnosti i srednie veka* [The history of natural science in ancient and medieval times]. St. Petersburg, 1909.
58. Tseyten, P. G. *Istoriya matematiki v drevnosti i srednie veka* [The history of mathematics in ancient and medieval times]. Moscow and Leningrad, 1938.
59. Heidegger, M. *Die Kategorien — und Bedeutungslehre des Duns Scottus*. Tübingen, 1916.
60. Darjes, I. G. *Via ad veritatem*, app., sec. 16, chap. 7 (In Logicam Lullii), pp. 263–275. Jena, 1764.
61. Roedt. *Commentarii in omnes parvos tractatus Parvorum logicalium Petri Hispani*. Cologne, 1493.
62. Prantl, Karl. *Geschichte der Logik im Abendlande*, vol. 4, sec. 18, pp. 145–146 (On Lully's "Art"). Leipzig, 1927.
63. Gardner, Martin. *Logic Machines and Diagrams*. New York, Toronto and London, 1958.
64. Lobstein. *Petrus Ramus*. Paris, 1878.
65. Herzen, Alexander. *Polnoye cobraniye cochineniy i pisem* [Complete works and letters], vol. 4, p. 120 (Remarks on Peter Ramus). St. Petersburg, 1912.
66. Cantor, M. "Petrus Ramus, ein wissenschaftlicher Märtyrer des sechzehnten Jahrhunderts." *Protestantische Monatsblätter von Gelzer*, vol. 30, 1867.
67. Desmaze, C. *Petrus Ramus, sa vie, ses écrits et sa mort*. Paris, 1864.
68. Darjes, I. G. *Via ad veritatem*, sec. 16, chap. 8 (In Logicam Rami), pp. 275–288. Jena, 1764.
69. Hausrath, A. *Petrus Abälard*. 1893.
70. Rémusat, Charles. *Abélard*, I, II. Paris, 1845.
71. Fedotov, G. *Abelyar* [Abélard]. Petrograd, 1924.

72. *Abaelardiana Inedita*. Logic of the 12th century: texts and commentaries, vol. 2. Rome, 1958.
73. Abelard, P. *Dialectica*. Edited by L. M. de Rijk. Assen, 1956.
74. Minio-Paluello, L. "The Ars Disserendi of Adam of Balsham, Parvipontanus." *Mediaeval and Renaissance Studies*, vol. III (1945).
75. Thomas, Ivo. "A Twelfth Century Paradox of the Infinite," *Journal of Symbolic Logic*, vol. 23, 1958.
76. Zubov, V. P. *Aristotel*. Moscow, 1963, pp. 236–239.
77. Lechler, G. V. *Robert Grosseteste, Bischof von Lincoln*, 1867 (Leipziger Universitätsprogramm).
78. ———. *Historisch-biographisches Handwörterbuch zur Geschichte der Philosophie* (Leipzig, 1879), s.v. "Robert Greathead."
79. Zubov, V. P. "Iz istorii srednevekovoy atomistiki" [On the history of medieval atomistics]. *Trudy Instituta istorii estestreznaniye i tekhniki*, vol. 1, 1947.
80. Minto, V. *Deduktivnaya i induktivnaya logika* [Deductive and inductive logic], pp. 18, 299, 331. Moscow, 1905.
81. Raydemeyster, K. "Doktor Mirabilis." *Znanie — sila*, no. 6, 1955.
82. Easton, S. C. *Roger Bacon and His Search for a Universal Science*. London, 1952.

B. Pascal (1623–1662)

83. Kol'man, E. "Znachenie simvolicheskoy logiki" [The meaning of symbolic logic]. In *Logicheskie issledovaniya*, 1959.
84. Scholz, H. "Pascal's demands of the mathematical method." *Festschrift zum 60. Geburtstag von Prof. Speiser*, Zurich, 1945. (German.)
85. Brunschweig. *Descartes et Pascal*. Paris, 1945.
86. Benze, Max. "The theory of science according to Blaise Pascal." *Zeitschrift für philosophische Forschungen*, vol. 2, no. 1, 1947. (German.)
87. Lescoeur, L. *Pascal's philosophical method*. (French) Dijon, 1850.
88. Fal'kenberg, R. *Istoriya novoy filosofii* [History of the new philosophy], p. 111. Moscow, 1910.
89. Reuchlin, F. H. *Pascal's Leben und der Geist seiner Schriften*. Paris, 1840.
90. Tavanets, P. V., and G. I. Ruzavin, "Osnovnye etapy razvitiya formal'noy logiki" [Basic stages in the development of formal logic]. In *Filosofskie voprosy sovremennoy formal'noy logiki* [Philosophical problems in contemporary formal logic], Moscow, 1962.
91. Asmus, V. F. *Dekart* [Descartes]. Moscow, 1962.

J. Jungius (1587–1657)

92. Meyer-Abich, A. "Joachim Jungius: ein Philosoph vor Leibniz." *Beiträge zur Leibnizforschung* I. 1947, pp. 138–152.
93. Wohlwill, E. *J. Jungius*. Hamburg, 1881.

A. Geulincx (1625–1669)

94. Dürr, Karl. Résumé of "Die mathematische Logik des Arnold Geulincx" *Journal of Symbolic Logic*, vol. 4, no. 4 (December 1939), p. 177.
95. *Historisch-philosophisches Handbuch zur Geschichte der Philosophie.* (Leipzig, 1879), s.v. "Geulincx, Arnold."

G. W. Leibniz (1646–1716)

96. Fischer, K. *Gottfried Wilhelm Leibniz. Leben, Werke und Lehre.* Heidelberg, 1902. [Russian translation: St. Petersburg, 1905.]
97. Schrecker, P. "Leibniz et le principe du tiers exclu," *Actes du Congrès International de Philosophie Scientifique*, vol. 6 (1936), pp. 75–84.
98. Dalbiez. "The fundamental idea of combinatorial analysis in Liebniz" (in French), *Travaux du IX Congres International de Philosophie*, V, VI. Reprinted in *Logic and Mathematics*, Paris, 1937, pp. 3–7.
99. Matzat, H. L. *Untersuchungen über die metaphysischen Grundlagen der Leibnizschen Zeichenkunst.* Berlin, 1938.
100. Wiener, Philip Paul. "Notes on Leibniz's Conception of Logic and its Historical Context." *The Philosophical Review*, vol. 48, 1939.
101. Quesada, F. M. "Sintesis de la conferencia sobre la Logica de Leibniz." *Actas de la Academía Nacional de Ciencias Exactas, Físicas y Naturales de Lima*, vol. 10, no. 1, 2 (1947), pp. 79–83.
102. Schischkoff, G. "Die gegenwärtige Logistik und Leibniz," *Beiträge zur Leibnizforschung* I, 1947, pp. 224–240.
103. Mahnke, D. "Leibnizens Synthese von Universalmathematik und Individualmetaphysik I," *Jahrbuch für Philosophie und phänomenologische Forschung* 7, September 1925.
104. Dürr, K. "Die mathematische Logik von Leibniz," *Studia Philosophica*, vol. 7 (1947), pp. 87–102.
105. ———. Leibniz' *Forschungen im Gebiet der Syllogistik.* Berlin, 1949.
106. Kauppi, R. "Über die Leibnizsche Logik, mit besonderer Berücksichtigung des Problems der Intension und der Extension," *Acta Philosophica Fennica*, XII, 1960.
107. Schrecker, P. "Leibniz and the Art of Inventing Algorithms," *Journal of the History of Ideas*, 8 (1947), 107–116.

108. Scholz, H. "Leibniz und die mathematische Grundlagenforschung," *Jahresbericht der Deutschen Mathematischen Vereinigung*, vol. 52, 1942, pp. 217–244.
109. Popov, P. S. *Istoriya logiki novogo opemeni* [History of recent logic], pp. 3, 22–30, 42, 52–122, 160, 198, 225, 239–246. Moscow, 1960.
110. Getmanova, A. D. "O vzglyadakh Leybnitsa na sootnoshenie matematik i logiki" [On Leibniz' views on the relationship between mathematics and logic]. In *Nekotorye filosofsko–teoreticheskie voprosy fiziki, matematiki i khimii* [Some philosophical and theoretical questions in physics, mathematics, and chemistry]. Moscow, 1959.
111. Venn, John. *Symbolic Logic*, pp. XXXI, 36, 52–57, 98, 146–165, 382–386, 407, 416–417. London, 1881.
112. Jorgensen, Jorgen. *Treatise of Formal Logic*, vol. I, pp. 15, 33, 58–59, 71–88, 117, 135, 147, 181, 193, 235; vol. III, pp. 19, 110, 129–130, 143, 158. Copenhagen and London, 1854.
113. Kvet, F. B. *Leibnizens Logik*. Prague, 1857.
114. Guhrauer, G. E. *Gottfried Wilhelm Freiherr von Leibniz*, 2 vols. Berlin, 1842.
115. Kabitz, W. *Die Philosophie des jungen Leibniz*. Heidelberg, 1909.
116. Tseyten, G. G. *Istoriya matematiki v XVI i XVII vv*. [The history of mathematics in the sixteenth and seventeenth centuries]. Moscow and Leningrad, 1938.
117. Russell, Bertrand. *A Critical Exposition of the Philosophy of Leibniz*. Cambridge, 1900.
118. Bykhovskiy, B. G. "Leybnits i ego mesto v istorii filosofiii" [Leibniz and his place in the history of philosophy]. Introduction to Leibniz' *New Experiments in Human Reason* (Russian translation), pp. 5–42. Moscow and Leningrad, 1936.
119. Korcik, Anthony. "Published and Unpublished Sources of Leibniz' Logic," *The Philosophical Review*, vol. 39, 1936.
120. Servien, Pius. "Le progrès de la métaphysique selon Leibniz et le langage des sciences," *Revue philosophique de la France et de l'Étranger*, vol. 124 (1937), pp. 140–154.
121. Huber, Kurt. "Leibniz and Us." *Zeitschrift für philosophische Forschung*, vol. 1, 1946–1947.
122. Lindemann, H. A. "Leibniz y la logica moderna," *Anales de la Sociedad Científica Argentina*, vol. 142 (1946), pp. 164–176.
123. Rescher, Nicolas. "Leibniz's Interpretation of his Logical Calculi." *The Journal of Symbolic Logic*, vol. 19, no. 1 (1954).
124. Zocher, Rudolf. "Zum Satz vom zureichenden Grunde bei Leibniz." *Beiträge zur Leibnizforschung*. 1947.
125. Hubscher, Arthur. "Leibniz und Schopenhauer." *Ibid.*, pp. 215–223.

126. Loemker, Leroy E. "A Note on the Origin of Leibniz's Discourse of 1686." *Journal of History of Ideas*. vol. 8, 1947, no. 4, pp. 449–466.
127. Wildon, Carr H. *Leibniz*. London, 1929.
128. Franco, Anerio. "Leibniz e Vico," *Giornale di Metafisica*, vol. 1, no. 6 (15 November, 1946), pp. 448–473.
129. Broad, C. D. "Leibniz's Last Controversy with the Newtonians," *Theoria*, vol. XII, no. 3, 1946, pp. 143–168.
130. Heinemann, Fritz. "Toland und Leibniz." *Beiträge zur Leibnizforschung*, 1947, pp. 193–214.
131. Viehweg, Theodor. "Die juristischen Beispielfälle in Leibnizens 'Ars Combinatoria.'" *Ibid.*, pp. 88–96.
132. Tönnies, F. "Leibniz und Hobbes." *Philosophische Monatshefte*, vol. XXIII, nos. 9 and 10 (1887), pp. 557–573.
133. Borighieri, P. *La disputa Leibniz-Newton sull'analisi*. Torino, 1958.
134. Herbert, Richard. *Die Lehre vom Unbewussten im System von Leibniz*. Halle, 1905.
135. Russell, B. "Recent work of the philosophy of Leibniz," *Mind*, April 1903.
136. Stein, Ludwig. *Leibniz und Spinoza*. Berlin, 1890.
137. Hartenstein, G. "Ueber Leibnizen's Lehre vom Verhältnis der Monaden zur Körperwelt." *Historisch-philosophische Abhandlungen*, 1870, pp. 469–537.
138. Rintelen, Fritz. "Leibnizen's Beziehung zur Scholastik." *AG. Ph.*, vol. 16, 1903.
139. Karinskiy, Vladimir. *Umozritel 'noe znanie v filosofskoy sisteme Leybnitsa* [Speculative knowledge in Leibniz' philosophical system]. St. Petersburg, 1912.
140. Brömse. *Das metaphysische Kausalproblem bei Leibniz*. Dissertation, Rostock, 1897.
141. Frenzel, B. *Der Associationsbegriff bei Leibniz*. Dissertation, Leipzig, 1893.
142. Hahn, Rud. *Die Entwicklung der Leibnizschen Metaphysik und der Einfluss der Mathematik auf dieselbe bis 1686*. Dissertation, Halle, 1898.
143. Pfleiderer, Edm. *Leibniz und Geulincx mit besonderer Beziehung auf ihr Uhrengleichnis*. Tübingen, 1884.
144. Vacca, Giovanni. "La logica di Leibniz," *Rivista di matematica*, v. 8. 1902–1906.
145. Dürr, Karl. "Neue Beleuchtung einer Theorie von Leibniz, Grundzüge des Logikkalküls." *Leibniz-Archiv*, no. 2. Darmstadt, 1930.
146. Vailati, Giovanni. "Sui carattere del contributio apportato dal Leibniz allo sviluppo della logica formale." *Rivista di filosofia e scienze affini*, vol. 12, 1905.

147. Couturat, Louis. *La logique de Leibniz d'après des documents inédits.* Paris, 1901.
148. Schmalenbach, H. *Leibniz.* Munich, 1921.
149. Matzat, Heinz-L. "Die Gedankenwelt des jungen Leibniz." *Beiträge zur Leibniz Forschung*, 1947, pp. 37–67.
150. De Carvalho, Joaquim. "Leibniz in seiner Beziehung zum portugiesischen Geistesleben." *Ibid.*, pp. 157–177.
151. De Careil, Foucher. *Nouvelles Lettres et Opuscules inédits de Leibniz.* Paris, 1857.
152. Cassirer, Ernst. *Leibniz's System in seinen wissenschaftlichen Grundlagen.* Marburg, 1902.
153. Ploucquet, Gottfried. "Anmerkungen über Leibnizens Difficultates Logicae." In *Sammlung der Schriften, welche den logischen Kalkül des Herrn Prof. Ploucqusts betreffen mit neuen Zusätzen*, ed. by Friedrich A. Böck. Tübingen, 1773.
154. Budylina, M. V. "Neizdannoe pis'mo Leybnitsa" [An unpublished letter of Leibniz]. Abstract in *Referativnyy Zhurnal Matemetika* [Journal of Mathematical Abstracts], 1955. Abstr. 4167.
155. Agostina. "Chetyre neizdannykh pis'ma Leybnitsa iodnopis'mo Guido Grandi" [Fourteen unpublished letters of Leibniz and one letter of Guido Grande]. *Ibid.*, 1955. Abstr. 2509.
156. Biermann, K. P. "Ob issledovanii G. V. Leybnitsem odnogo voprosa kombinatoriki" [On the investigation by Leibniz of a question in combinatorial analysis]. *Ibid.*, 1958. Abstr. 893.
157. ———. "'Ob otsenke nedostorernogo' Leybnitsa" [Leibniz' "On investigating the uncertain"]. *Ibid.*, 1958. Abstr. 5372.
158. Resher, N. "Leybnitsevskaya kontseptsiya kolichestra, chisla i beskonechnosti" [The Leibnizian conception of quantity, number, and infinity]. *Ibid.*, 1956. Abstr. 70.
159. Biermann, K. P. Ob odnoy zametke Leybnitsa po voprosam teorii veroyatnostey [On a note of Leibniz on questions in the theory of probability]. *Ibid.*, 1956. Abstr. 3571.
160. Vailati, Giovanni. "Review of Couturat's La logique de Leibniz." *Rivista di matematica*, vol. 7 (1900–1901).
161. Fehr, H. "Review of Couturat's 'La logique de Leibniz.'" *L'Enseignement mathématique*, vol. 5 (1903).
162. Duncan, G. M. "Review of Couturat's 'La logique de Leibniz.'" *The Philosophical Review*, vol. 12, 1903.
163. Jourdain, P. E. B. "The Logical Work of Leibniz," *The Monist*, vol. 26, 1916.
164. Russell, B. *History of Western Philosophy* (Russian translation). Moscow, 1960, p. 613 ff.
165. Couturat, Louis. *Opuscules et fragments inédits de Leibniz extraits des manuscripts de la Bibliothèque royale de Hanovre.* Paris, 1903.

166. God. Gull. Leibnitii. *Opera philosophica quae extant latina, gallica, germanica omnia*. Edited by J. E. Erdmann. Berlin, 1840.
167. ———. *Leibnizens mathematische Schriften*. Edited by C. J. Gerhardt, 7 vols. Berlin, 1849–1863.
168. ———. *Die philosophischen Schriften von Gottfried Wilhelm Leibniz*. Edited by C. J. Gerhardt. Berlin, 1875–1890.
169. ———. *Selected Philosophical Works* (Russian translation). 4th ed. Moscow, 1890.
170. ———. *Philosophische Schriften, herausgegeben von der Preussischen Akademie der Wissenschaft*, vol. 1. Darmstadt, 1930.
171. ———. *Philosophische Werke nach Raspens Sammlung*. Edited by J. H. F. Ulrich. Halle, 1780.
172. C. J. Gerhardt's *Philosophische Schriften*. Berlin, 1875–1890.
173. ———. *Selected Works*. New York, 1951.
174. ———. *New Experiments in Human Reason* (Russian translation). Moscow and Leningrad, 1936.
175. Feuerbach, L. *Darstellung, Entwicklung und Kritik der Leibniz'schen Philosophie*. Berlin, 1837.
176. Janet, P. *Oeuvres philosophiques de Leibniz, avec une introduction et des notes*. 2 vols. 1866.

Jakob Bernoulli (1654–1705); Johann Bernoulli (1667–1748)

177. Bernoulli, Jacques and Jean Bernoulli. *Parallelismus ratiocini logici et algebraici*. Basel, 1685. Reprinted in *The Works of Jakob Bernoulli of Basel*, vol. I, pp. 211–224. Geneva, 1744.
178. ———. *Journal of Symbolic Logic*, vol. 1 (1936), p. 125.
179. Glivenko, V. I. "Osnovy obshchey teorii struktur" [Fundamentals of the general theory of structures]. *Ucheniye zapiski Moskovskogo pedagogicheskoy instituta, seriya fiziko–matematicheskaya*, vol. 1, 1937.
180. *Der Briefwechsel von Johann Bernoulli*, vol. 1. Basel, 1955.
181. Vipper, Yu. F. *Semeystvo matematikov Bernulli* [The Bernoulli family of mathematicians]. Moscow, 1875.
182. Fleckenstein, J. *Johann und Jakob Bernoulli*. Basel, 1949.

Christian von Wolff (1679–1754)

183. De Crousaz, Jean Pierre. *Observations critiques sur l'abrégé de la logique de Mr. Wolff*. 1744.
184. Kluge, F. W. *Christian von Wolff, der Philosoph*. 1831.
185. Schlösser, F. P. *Disputatio de sororio Logices et Matheos nexu, et applicatione praeceptorum logicorum in disciplinis mathematicis*. 1727.
186. *Historisch-biographisches Handwörterbuch zur Geschichte der Philosophie* (Leipzig, 1879), s.v. "Wolff, Christian."

187. Pichler, H. *Über Chr. Wolffs Ontologie.* Leipzig, 1910.
188. Baumann, J. *Wolffsche Begriffsbestimmungen.* Leipzig, 1910.
189. Utitz, E. *Chr. Wolff.* Hale, 1929.
190. Campo, M. *Chr. Wolff e il razionalismo precritico.* Milano, 1939.
191. Mühlprfordt, G. "Christian Wolff, ein Enzyklopädist der deutschen Aufklärung." In *Jahrbuch für Geschichte der deutsch-slawischen Beziehungen*, vol. 1. Halle, 1956.
192. *Filosofskaya entsiklopediya* [Philosophical encyclopedia] (Moscow, 1960), s.v. "Kh. Vol'f" (C. Wolff), by B. Rabbot.
193. Wolff, Christian. *Razumnye mysli o silakh chelovecheskogo rassudka* [Rational thoughts on the powers of the human intellect]. St. Petersburg, 1765.
194. ———. *Vernünftige Gedanken von den Kräften des menschlichen Vestandes.* Magdeburg, 1754.
195. ———. *Psychologia empirica methodo scientifica pertractata . . .* Frankfurt and Leipzig, 1732.
196. Baumgarten, Alexander Gottlieb. "Acroasis logica." In *C. L. B. De Wolff.* Halles, 1761.
197. Venn, John. *Symbolic Logic*, pp. XXXI, XXXVI, 98, 443. London, 1881.
198. Wolff, Christian. *Philosophia rationalis sive Logica.* Frankfurt and Leipzig, 1732.
199. ———. *Sokrashchenie pervykh osnovaniy matematiki* [Abridged first principles of mathematics], vol. 1, St. Petersburg, 1770.

J. A. Segner (1704–1777)

200. Segner, J. A. *Specimen logicae universaliter demonstratae.* Jena, 1740.
201. Venn, John. *Symbolic Logic*, pp. XXXI, 95, 184, 407, 412–413, 442. London, 1881.
202. ———. "On the various notations adopted for expressing the common propositions in logic" (read 6 December 1880). *Proceedings of the Cambridge Philosophical Society*, vol. 4 (1880–1883), pp. 36–47.
203. Vailati Giovanni. "La logique mathématique et sa nouvelle phase de développement dans les écrits de M. J. Peano." *Revue de métaphysique et de Morale*, vol. 7 (1899).

G. Ploucquet (1716–1790)

204. Aner, Karl. "Gottfried Ploucquets Leben und Lehren." *Abhandlungen zur Philosophie und ihre Geschichte*, vol. 33. Halle, 1909.
205. Ploucquet, Gottfried. *Expositiones philosophiae theoreticae.* Stuttgart, 1782.

206. Glemm, H. W. "Ploucquet G." *Novae amoenitates litterarie*, vol. 4 (1764).
207. Abbt, Thomas. "Auszug aus Hrn. Pr. Ploucquets 'Methodo calculandi in Logicis.'" *Briefe die neueste Literatur betreffend*, vol. 17, 1764.
208. ———. "Anmerkungen über diesen Methodum," *Ibid.*, pp. 81–94.
209. ———. "Beschluss dieser Anmerkungen," *Ibid.*, pp. 95–104.
210. *Historisch-biographisches Handwörterbuch zur Geschichte der Philosophie* (Leipzig, 1879), s.v. "Ploucquet."
211. Bornstein, P. *Gottfried Ploucquet's 'Erkenntnistheorie und Metaphysik.'* Dissertation, Potsdam, 1898.
212. Venn, John. *Symbolic Logic*, pp. 382, 402–430, 442. London, 1881.
213. *Dictionnaire des sciences philosophiques* (Paris, 1885), s.v. "Godefroy Ploucquet."
214. Ebernstein, V. *Histoire de la logique et de la métaphysique*, vol. 1. Paris, 1880.
215. Hegel. *Sochineniya* [Works] (Russian translation), Vol. 6, Moscow, 1939.
216. Erdmann, B. *Logik*, 1st ed., 1892, 2nd ed., 1907.
217. Lambert, J. H. *Neue Zeitungen von gelehrten Sachen*, nos. 1, 58, 59. Leipzig, 1765.
218. Ploucquet, G. *Sammlung der Schriften, welche den logischen Kalkül des Herrn Prof. Ploucquets betreffen mit neuen Zusätzen.* Tübingen, 1773.

J. H. Lambert (1728–1777)

219. König, E. "Begriff der Objektivität bei Wolff und Lambert," *Zeitschrift Philosophie Kr.*, vol. 84, 1884.
220. Dürr, Karl. *Die Logistik J. H. Lamberts.* Zurich, 1945.
221. Eisenring, M. E. *Johann Heinrich Lambert und die wissenschaftliche Philosophie der Gegenwart.* Dissertation, Zurich, 1944.
222. Krienelke, Karl. *J. H. Lambert's Philosophie der Mathematik.* Dissertation, Halle, 1909.
223. *Briefwechsel zwischen Leonhard Euler und J. H. Lambert.* Berlin, 1924.
224. Baensch, Otto. *Johann Heinrich Lamberts Philosophie und seine Stellung zu Kant.* Tübingen and Leipzig, 1902.
225. Jorgensen, Jorgen. *Treatise of Formal Logic*, vol. 1, pp. 1, 6, 33, 82–97; vol. 3, p. 258. Copenhagen and London, 1931.
226. Venn, John. *Symbolic Logic*, pp. XXXI–XXXVIII, 8, 36, 54, 78, 99, 157, 195, 272–274, 321, 328, 358, 382–383, 398–411, 441. London, 1881.

227. Ploucquet, G. *Antwort auf die von Herrn Professor Lambert gemachten Erinnerungen und diesseitiger Beschluss der logicalischen Rechnungsstreitigkeit.* Leipzig, 1766.
228. ———. *Untersuchung und Abänderung der logikalischen Konstruktionen des Herrn Professors Lambert.* Tübingen, 1765.
229. *Historisch-biographisches Handwörterbuch zur Geschichte der Philosophie* (Leipzig, 1879) s.v. "Lambert."
230. Lambert, J. H. *Beiträge zum Gebrauch der Mathematik und deren Anwendung*, vols. 1–3. Berlin, 1765–1772.
231. ———. *Opera mathematica*, vol. 1. Turici, Füs. Cop. 1946.
232. ———. Examen d'une espèce de superstitution ramené au calcul des probabilités." *Mémoires de l'Académie de Berlin*, 1771.
233. *Dictionaire des sciences philosophiques* (Paris, 1885), s.v. "Lambert."
234. Styazhkin, N. I. Reference 324, pp. 96–97.
235. Lambert, J. H. *Kosmologische Briefe.* Leipzig, 1761.
236. *B.S.E.*, 2nd ed., s.v. "Lambert."
237. Barthel, E. "Johann Heinrich Lambert." *Archiv für Geschichte der Mathematik, der Naturwissenschaft und der Technik*, vol. 11, nos. 1, 2. Berlin, 1829.
238. Lambert, J. H. *Logische und philosophische Abhandlungen.* 2 vols. Dessau, 1782–1787.
239. ———. *Neues Organon*, vol. 2, sec. 1, Semiotics. Leipzig, 1764.
240. ———. *Anlage zur Architectonic oder Theorie des Einfachen und des Ersten in der philosophischen und mathematischen Erkenntniss*, vols. 1, 2. Riga, 1771.
241. ———. *Deutscher Gelehrter Briefwechsel.* 4 vols. Berlin, 1781–1784.
242. Marx, Karl. *K kritike politicheskoy ekonomii* [Toward a critique of political economy] (Russian Edition, p. 46). Moscow, 1952.

S. Maimon (1753–1800)

243. Maimon, Solomon. *Versuch einer neuen Logik oder Theorie des Denkens nebst angehängten Briefen des Philaletes an Aenesidemus von Solomon Maimon.* Berlin, 1794; Berlin, 1912. [In this work we have cited the 1912 edition].
244. Engel, Bernard Carl. *Maimons Leben.* Berlin, 1912. Introduction to the Berlin edition, 1912, cited in Reference 243.
245. Solomon Maimons Lebensgeschichte. *Von ihm selbst geschrieben und herausgegeben* von K. Ph. Moritz, 1792–1793. 2 vols.
246. Gottselig. "Die Logik S. Maimons." *Berner Studien*, 1908.
247. Rubin. "Die Erkenntnistheorie Maimons in ihrem Verhältnis zu Cartesius, Leibniz, Hume und Kant." *Berner Studien zur Philosophie und ihrer Geschichte*, vol. 7, 1879.
248. Witte, John H. *Solomon Maimon.* Berlin, 1876.

249. *Historisch-biographisches Handwörterbuch zur Geschichte der Philosophie.* (Leipzig, 1879), s.v. "Solomon Maimon."
250. Wolff. *S. J. Maimoniana oder Rhapsodien zur Characteristik Sol. Maimons, aus seinem Privatleben gesammelt.* Heidelberg, 1814.
251. Kuntze, Friedrich. *Die Philosophie S. Maimons.* Heidelberg, 1912.
252. *Evreyskaya biblioteka* [*The Jewish Library*], vols. 1, 2. St. Petersburg, 1871–1872.
253. Venn, John. *Symbolic Logic*, pp. XXXVI, 52–54, 94, 377, 407, 420, 441. London, 1881.
254. Bergmann, Hugo. *Maimons logischer Kalkül*, *Philosophia*, vol. 3. Belgrade, 1938.

G. F. Castillon (1747–1814)

255. Castillon, F. "Memoire sur un nouvel algorithme logique." *Mémoires de l'Académie Royale des Sciences et Belles-Lettres*, vol. 53, 1805. Classe philosophie spéculative.
256. Venn, John. "Notice of Castillon's: Sur un nouvel algorithme logique." *Mind*, vol. 6, 1881.
257. Lewis, C. I. *A Survey of Symbolic Logic.* Berkeley, 1918.

C. A. Semler

258. Semler, C. August. *Versuch über die combinatorische Methode: ein Beitrag zur angewandten Logik und allgemeinen Methodik.* Dresden, 1811.

J. D. Gergonne (1771–1859)

259. Vacca, Giovanni. "Sui precursori della logica mathematica II. J. D. Gergonne." *Rivista ai matematica*, vol. 6, 1896–1899.

A. D. Twesten (1789–1876)

260. Twesten, A. D. *Die Logik, insbesondere die Analytik.* Schlesweig, 1825.
261. ———. *Grundriss der analytischen Logik.* Kiel, 1834.
262. Leikfeld, P. Reference 337.
263. Subbotin, A. L. "O tsepyakh klassicheskikh sillogizmov" [On chains of classical syllogisms]. *Nauchnye doklady vysshey shkoly*, no. 3, 1959.

K. F. Gauber (1775–1851)

264. Gauber, K. F. *Scholae logico-mathematicae.* Stuttgart, 1829.

B. Bolzano (1781–1848)

265. Dubislaw, W. "Bolzano als Vorläufer der mathematischen Logik." In *Philosophisches Jahrbuch der Gömes-Gesellschaft*, vol. 44, 1931.
266. Bolzano, Bernard. *Wissenschaftslehre*. 4 vols. Sulzbach, 1837.
267. Kol'man, E. Ya. *Bernard Bol'tsano* [Bernard Bolzano]. 1955.
268. Church, A. "Propositions and sentences." In *The Problem of Universals*. Notre Dame, South Bend, Ind., 1956.
269. Husserl, Edmund. *Logische Untersuchungen*. Halle, 1901.
270. Venn, John. *Symbolic Logic*, pp. 426–427, London, 1881.
271. Korselt, Alwin. *Review of Bolzano's Wissenschaftslehre*, vol. 1 (1914); *Jahresbericht D.M.—V.*, section 2, vols. 2, 3. 1914–1915.
272. Nutsubidze, Sh. "Bol'tsano i teoriya nauki" [Bolzano and the theory of science], *Voprosy fiilosofii i psikhologii*, vol. 116–117, 1913.
273. *Historisch-biographisches Handwörterbuch zur Geschichte der Philosophie*. (Leipzig, 1879), s.v. "Bolzano, B."
274. Heesch, E. "Grundzüge der Bolzanoschen Wissenschaftslehre." *Philosophie Jahrbuch der Görres-Gesellschaft*, vol. 48, 1935.
275. Scholz, Heinrich. "Die Wissenschaftslehre Bolzanos" in *Semesterberichte* (Münster), 9 Sem., Winter, 1936–1937. Erweit, in *Abhandl. d. Fries'schen Schule*, no. 6.
276. Smart, H. "Bolzano's Logic." *The Philosophical Review*, vol. 53, 1944.
277. Theobald, S. *Die Bedeutung der Mathematik für die logischen Untersuchungen Bernard Bolzanos*. Dissertation, Bonn, 1929, Koblenz, 1928.
278. Bar–Hillel, Y. "Bolzano's definition of analytical propositions." *Methodos*, vol. 2, 1950.
279. Gladkiy, A. V. Abstract 2680. *Referativnyy Zhurnal Matemetika*, no. 3, (1960), pp. 10–11.
280. Bakradze, K. S. *Ocherki po istorii noveyshey i sovremennoy burzhuaznoy filosofiya* [Notes on the history of recent and contemporary bourgeois philosophy], pp. 436–440, Tbilisi, 1960.
281. Beth, E. W. *The philosophy of mathematics from Parmenides to Bolzano*. Nimèque, 1944. (French).
282. Winter, E. J. *Leben und geistige Entwicklung des sozialethikers und mathematikers Bernard Bolzano* (1781–1848). Halle, 1949.
283. ———. "Der böhmische Vormärz in Briefen B. Bolzanos an Přihonsky (1824–1848)." *Beitrage zur deutsch-slawischen Wechselseitigkeit*, Berlin.
284. Grushkova, T. "Referat stat'i Rychlik Karel," Bolzanuv pobyt v Libechové [Review of an article on Bolzano]. *Matematicheskoy skole*, vol. 9, no. 2 (1959).

285. Bolzano, B. *Paradoksy beskonechnogo* [Paradoxes of the infinite]. Odessa, 1911.
286. Vileytner, G. *Istoriya matematiki ot Dekarta do serediny XIX veka* [The history of mathematics from Descartes to the middle of the nineteenth century]. Moscow, 1960.

G. Bentham (1800–1884)

287. Wisdom, J. *Interpretation and Analysis in Relation to Bentham's Theory of Definition.* London, 1931.
288. Bentham, George, *Outline of a New System of Logic.* London, 1827.

W. Hamilton (1788–1856)

289. *Brokgauz & Evfron Encyclopedic Dictionary* (St. Petersburg, 1892), s.v. "Hamilton," by E. L. Radlov.
290. Hamilton, W. *Lectures on Metaphysics and Logic,* vols. 1–4. Edinburgh and London, 1860.
291. ———. "Philosophy of Unconditioned." *Edinburgh Review,* 1829.
292. Bourdillat, Francois. *La réforme logique de Hamilton.* Paris, 1891.
293. Mill, J. S. *Obzor filosofii sera Vil'yama Gamil'tona i glavnykh filosofskikh voprosov, obsuzhdennykh vego tvoreniyakh* [Review of the philosophy of Sir William Hamilton and the principal philosophical questions discussed in his works]. St. Petersburg, 1869.
294. Tsutsugan, F. "Critica teoriei cautificarii predicatului, a lui W. Hamilton." *Revista de filozofie,* nos. 1, 2, 1943.

M. W. Drobisch (1802–1896)

295. Drobisch, M. W. *Neue Darstellung der Logic nach ihren einfachen Verhältnissen nebst einem logisch-mathematischen Anhange.* Leipzig, 1872.
296. Bransch, Moritz. *Leipziger Philosophen.* Leipzig, 1894.
297. Credaro, Luigi, "Maur. Guilgelmo Drobisch." *Rivista Italiana di filosofia,* Rome, 1897.

A. De Morgan (1806–1871)

298. *Brokgauz & Evfron Encyclopedic Dictionary* (St. Petersburg, 1892), s.v. "De Morgan," by V. V. Bobynin.
299. De Morgan, Augustus. *Syllabus of a Proposed System of Logic.* 1860.
300. "Recent extensions of formal logic. A Review of De Morgan's 'Formal Logic' and 'On the symbolic logic'; Thompson's

'Outline of the necessary laws of thought,' and Bayne's 'Essay on the new analytic of logical forms.'" *The North British Review*, vol. 15, 1851; American edition, vol. 10, 1851, pp. 47–63.
301. Liard, L. *Les logiciens anglais contemporains*. Paris, 1907.
302. De Morgan, Sophie E. *Memoirs of Augustus De Morgan*. London, 1882.
303. Halsted, G. B. "De Morgan as Logician." *Journal of Speculative Philosophy*, vol. 18, 1884.
304. Troitskiy, M. M. *Uchebnik logiki s podrobnymi ukazaniyami na istoriyu i sostoyanie etoy nauki v Rossii i drugikh stranakh* [Handbook of logic with detailed notes on the history and state of this science in Russia and in other countries], pp. 206–207. Moscow, 1886.
305. De Morgan, A. *Formal Logic, or the Calculus of Inference, Necessary and Probable*. London, 1847.
306. Styazhkin, N. I. Reference 324, p. 98.

G. Boole (1815–1864)

307. Boole G. *Collected logical works*, vol. 2. Chicago and London, 1940.
308. ———. *Studies in logic and probability*, vol. 1. London, 1952.
309. Rhees, R. Note in editing of *Collected Logical Works of G. Boole*. Chicago and London, 1940.
310. Scholz, H. *Zentralblatt für Mathematik und ihre Grenzgebiete*, vol. 4, no. 1/3, September, 1954.
311. Harley, R. "Remarks on Boole's Mathematical Analysis of Logic." *Report of the 36th meeting of the British Association for Advancement of Science*, (1867).
312. ———. *Report of 40th meeting of the British Association for Advancement of Science*. London, pp. 14–15.
313. Couturat, L. *Algebra logiki* [The algebra of logic], pp. 65–67. Odessa, 1909.
314. Poretskiy, P. S. "Sept lois fondamentales de la thèorie des égalités logiques." *Izvestiya fiziko-matematicheskogo obshchestva pri Kazanskom universitete*, vol. 8, 1898.
315. Zhegalkin, I. I. "Arifmetizatsiya simbolicheskoy logiki" [Arithmetization of symbolic logic]. *Mat. sb.*, vol. 35, 1928.
316. Poretskiy, P. S. *O sposobakh resheniya logicheskikh ravenstv i ob obratnom sposobe matematicheskoy logiki* [On methods of solution of logical equations and on the inverse method in mathematical logic]. Kazan, 1884.
317. Lotze, H. *Logik*. Leipzig, 1880.
318. Kozlovskiy, F. "Simvolicheskiy analiz form i protsessov mysli, sostablyayushchikh predmet formal 'noy logiki'" [Symbolic

analysis of the forms and processes of thought which comprise the subject of formal logic]. *Kievskie universitetskie izvestiya*, nos. 1, 2 (1882).

319. Venn, John. (Article on Boole.) *Mind* vol. 1 (1876), pp. 481–485.
320. Bryant, Sophie. "On the Nature and Functions of a Complete Symbolic Language." *Mind*, vol. 13 (1888).
321. Poretskiy, P. S. "Zakon korney v logike" [The law of roots in logic]. *Nauchnoe obozrenie*, no. 19 (1896).
322. Feys, R. *Proceedings of the Royal Irish Academy*, vol. 57, no. 6 (1955).
323. Stone, M. (Article on Boolean rings.) *Proceedings of the National Academy of Science*, vol. 21 (1935), pp. 103–105.
324. Styazhkin, N. I. "K kharakteristike ranney stadii v razvitii idey matematicheskoy logiki" [Toward a characterization of the early stages of the development of mathematical logic]. *Nauchnye doklady vysshey shkoly: Filosofskie nauki*, no. 3 (1958), pp. 100–101.
325. Schröder, E. *Der Operationskreis des Logikkalküls*. Leipzig, 1877.
326. Jorgensen, Jorgen, *Treatise of formal logic*, vol. 1, pp. 1, 33, 70–201. Copenhagen and London, 1931.
327. Boole, G. *An Investigation of the Laws of Thought on Which Are Founded the Mathematical Theories of Logic and Probabilities*. London, 1854.
328. Venn, John. *Symbolic Logic*. London, 1881; London and New York, 1894.
329. Styazhkin, N. I. *Iz istorii razvitiya matematicheskoy logiki v XIX v.* [On the history of the development of mathematical logic in the nineteenth century]. Dissertation, 1958.
330. Macfarlane, A. *Lectures on ten British mathematicians*, p. 50. 1916.
331. Birkhoff, G. *Teoriya struktur* [The theory of structures]. Moscow, 1952. (For Boolean algebras, see pp. 215–217, 229–234, 243 of Russian edition.)
332. Thomas, Ivo. "Boole's concept of science." *Proceedings of the Royal Irish Academy*, vol. 57, sec. B, no. 6 (1955).
333. Sleszyński, Jan. *Teoria dowodu*, vol. 2. Krakow, 1929.
334. Hesse, Mary B. "Boole's philosophy of logic." *Annales of Science*, vol. 8, 1952.
335. Liard, L. *Angliyski reformatory logiki v XIX v.* [English reformers of logic in the nineteenth century]. St. Petersburg, 1897.
336. ———. *Les logiciens anglais contemporains*. Paris, 1878.
337. Leikfeld, P. *Razlichnye napravleniya v logike i osnovyne zadachi etoy nauki* [Various directions in logic and the basic problems of this science]. Kharkov, 1890.
338. Hefler, Aloise. *Osnovnye ucheniya logiki* [Basic studies in logic]. St. Petersburg, 1910.

339. Russell, L. J. *Philosophy*, vol. 29, 1954.
340. Feys, Robert. "Boole as a Logician." *Proceedings of the Royal Irish Academy*, vol. 57, sec. B, no. 6 (1955).
341. Rosser, J. B. "Boole and the Concept of a Function." *Ibid.*
342. Rhees, R. G. "Boole as Student and Teacher." *ibid.*
343. Hackett, F. E. "The Method of George Boole." *ibid.*
344. Russell, Bertrand. "Recent Work on the Principles of Mathematics." *The International Monthly*, vol. 4, no. 1, 1901.
345. Poretskiy, P. S. "Reshenie obshchey zadachi teorii veroyatnostey pri promoshchi matematicheskoy logiki" [Solution of general problems in the theory of probability by using mathematical logic]. *Sobranie protokolov sektsii obshchestva estestvoispytateley prirody pri Kazanskom universitete*, vol. 5, September, 1886–May 1887.
346. Taylor, Geoffrey. "George Boole." *Proceedings of the Royal Irish Academy*, vol. 57, sec. A, no. 6. *Celebration of the centenary of the "Laws of Thought," by George Boole*, pp. 66–73. Dublin, 1955.
347. Boole, Mary. "Home Side of a Scientific Mind." *The University Magazine*, 1878, pp. 105, 173, 326, 454.
348. Harley, R. "George Boole." *The British Quarterly Review*, vol. 44, 1866.
349. Anonymous. "George Boole." *Proceedings of the Royal Society of London*, vol. 15 (1867).
350. Hughlings, J. P. *The Logic of Names, an Introduction to Boole's 'Laws of Thought,'* London, 1869.
351. Halsted, G. B. "Professor Jevons' Criticism of Boole's Logical System." *Mind*, vol. 3, 1878.
352. Young, G. P. "Remarks on Professor Boole's Mathematical Theory of Laws of Thought," *The Canadian Journal of Science and History*, vol. 10, Toronto, 1865.
353. *The Encyclopaedia Britannica* (1953), s.v. "George Boole," by W. S. Jevons.
354. *Dictionary of National Biography* (London and New York, 1886), s.v. "George Boole," by John Venn.
355. Bobynin, V. V. *Opyty matematicheskogo izlozheniya logiki* [Investigations in the mathematical exposition of logic]. 1st ed., Moscow, 1886. (The work of G. Boole and R. Grassmann.)
356. Klein–Barmen, Fritz. "Boole–Schröder Structures." *Deutsche Mathematik*, vol. 1, 1936.
357. Harley, R. "A Contribution to the History of the Algebra of Logic." *Report of the Fiftieth meeting of the British Association for the Advancement of Science*, London, 1881.
358. Nagel, E. "Impossible Numbers: a Chapter on the History of Modern Logic," in *Studies in the History of Ideas*. New York, 1935.
359. Prior, A. N. "Categoricals and Hypotheticals in George Boole

and His Successors." *The Australian Journal of Philosophy*, vol. 27, 1949.
360. Hilbert, D., and W. Ackermann. *Grundzüge der theoretischen Logik*. Berlin, 1928. (Russian translation; Moscow, 1947.)
361. Kneale, William. "Boole and the Revival of Logic." *Mind*, vol. 57 (1948), pp. 149–175.
362. Murphy, J. J. "On the Quantification of Predicates and on the Interpretation of Boole's Logical Symbols." *Proceedings of the Literary and Philosophical Society of Manchester*, vol. 23, (1884), pp. 33–36.
363. Styazhkin, N. I. "Obosnovanie i analiz logicheskikh metodov Dzhordzha Bulya" [Verification and analysis of the logical methods of George Boole]. *Vestnik MGU*, seriya VIII (ekonomiki i filosofii), no. 1, 1960.
364. Lewis, C. I. *Survey of Symbolic Logic*. Berkeley, 1918.
365. Liard, L. "La logique algébrique de Boole." *Revue Philosophique*, vol. 4, 1877.
366. Frege, Gottlob. "Über die Begriffsschrift des Herrn Peano und meine eigene." *Berichte der mathematischen Classe der königlich Sächsischen Akademie der Wissenschaften zu Leipzig*, 1896.
367. Ellis, A. J. "On the Algebraic Analogues of Logical Relations." *Proceedings of the Royal Society of London*, vol. 21 (1873), pp. 497–498.
368. Popov, P. S. Reference 109, pp. 218–220.
369. Tigert, Ino I. *Hand-book of logic*. Nashville, 1886.
370. Bochenski, J. *Formale logik*. Munich, 1956.

W. S. Jevons (1835–1882)

371. Mays, W., and D. P. Henry. Exhibition of the Works of W. Stanley Jevons. *Nature*, vol. 170, 1952.
372. Adamson, R. "Professor Jevons in Mill's Experimental Methods." *Mind*, vol. 3, 1878.
373. Orlov, S. "Novaya sistema formal'noy logiki" [A new system of formal logic]. (A review of S. Jevons's book *The Principles of Science*.) *Zhurnal Ministerstva Narodnogo Prosveshcheniya*, pt. 217, 1881.
374. Sleshinskiy, I. V. "Logicheskaya mashina S. Dzhevons" [S. Jevons' logical machine]. *Vestnik opytnoy fiziki i elementarnoy matematiki*, 1893, no. 175 and no. 7, pp. 145–154.
375. Robertson, G. C. "Review of Jevons's *The Principles of Science*." *Mind*, vol. 1 (1876).
376. Mays, W., and D. P. Henry. "Jevons and Logic." *Mind*, vol. 62 (1953).

377. Adamson, Robert. "Review of Jevons's *Studies in Deductive Logic.*" *Mind*, vol. 6 (1881).
378. *Encyclopaedia Britannica.* (1902), s.v. "W. S. Jevons," by J. N. Keynes.
379. Liard, L. "Un nouvel systéme de logique formelle W. Stanley Jevons." *Revue philosophique de la France et de l'étranger*, vol. 3, 1877.
380. Jevons, W. S. "On the Mechanical Performance of Logical Inference." *Philosophical Transactions*, 1870, vol. 160; pt. 2.

E. Schröder (1841–1902)

381. Schröder, E. *Der Operationskreis des Logikkalküls.* Leipzig, 1877.
382. Church, Alonzo. "Schröder's Anticipation of the Simple Theory of Types." *Journal of Symbolic Logic*, vol. 4, no. 4 (December 1939).
383. Schröder, Ernst. *Vorlesungen über die Algebra der Logic*, vols. 1, 2. Leipzig, 1890–1891.
384. Behrens, Georg. *Die Prinzipien der mathematischen Logic bei Schröder, Russell and König.* Dissertation, Hamburg, 1918.
385. Bobynin, V. V. *Opyty matematicheskogo izlozheniya logiki. Raboty E. Schrodera* [Experiments in the mathematical presentation of logic. The work of E. Schröder]. 2nd ed. Moscow, 1894.
386. Jorgensen, Jorgen. Reference 326, vol. I, pp. 46, 81, 88, 124–137, 147, 175, 253–261; vol. 2, pp. 20, 41–43, 205, 262.
387. Lüroth, J. "Ernst Schröder." *Jahresbericht der Deutschen Mathematischen Vereinigung*, vol. 12, 1903.
388. Reyes y Prosper, V. "Ernesto Schroeder. Sus merecimientos ante la Logica, su propaganda logico-matematica, sus obras." *El progresso matemático*, 2, 1892.
389. Pereira, Ramón Crespo. "On Schröder's algebra of logic." (Spanish.) *Revista matemática hispano-americana*, series 4, vol. 11, 1951.
390. *Filosofskaya entsiklopediya*, s.v. "Dzhevons, U.S." [W. S. Jevons] by B. V. Biryukov.
391. Klein Bermen, Fritz. Reference 356.
392. Schröder, E. *Lehrbuch der Arithmetik und Algebra*, vol. 1, 1873.
393. Stringham, W. J. "Review of Schröder's 'Der Operationskreis des Logikkalküls.'" *John Hopkins University Circulars*, vol. 1, (1879–1882), p. 49.

H. McColl (1837–1909)

394. McColl, Hugh. Symbolic or Abbreviated Language, with Application to Mathematical Probability." *Mathematical Quaestiones*, vol. 28 (1877), pp. 20–23.

395. McCoil, Hugh. "The Calculus of Equivalent Statements and Integration Limits." *Proceedings of the London Mathematical Society*, vol. 9 (1877–1878), vol. 10 (1878–1879), vol. 11 (1880).
396. ———. "'If' and 'imply.'" *Mind*, n. s., vol. 17 (1908), pp. 151–152; 453–454.

P. S. Poretskiy (1846–1904)

397. Blake, A. *Canonical Expressions in Boolean Algebra*. Chicago, 1938.
398. McKinsey, J. C. "Review of *Canonical Expressions in Boolean Algebra*." *Journal of Symbolic Logic*, vol. 3, no. 2 (1938).
399. Yagodinskiy, I. I. "Geneticheskiy metod v logike" [The genetic method in logic]. *Ucheniye zapiski Kazanskogo universiteta*, 1909, p. 14.
400. Dubyago, D. I. "P. S. Poretskiy" (obituary notice), *Izvestiya fizichesko matematicheskogo obshchestva pri Kazanskom universitete*, 2nd ser., vol. 16, no. 1 (1908), p. 6.
401. Poretskiy, P. S. Reference 316 in *Sobranie protokolov zasedaniy sektsii fizichesko matematicheskikh nauk pri Kazanskom universitete*, vol. 2, Kazan, 1884.
402. ———. Reference 345.
403. ———. Reference 321.
404. ———. "Sept lois fondamentales de la théorie des égalités logiques," *Izvestiya fizichesko matematicheskogo obshchestva pri Kazanskom universitete*, sec. 2, nos. 1–3; vol. 9, no. 2, 1900–1901.
405. ———. "Quelques lois ultérieures de la théorie des égalités logiques." *Ibid.*, sec. 2, nos. 1–3, 12, 1900–1901.
406. ———. "Appendice sur mon nouvel travail: Théorie des non-égalités logiques." *Ibid.*, vol. 14, no. 2, 1904.
407. ———. "Théorie conjointe des égalités et des non-égalités logiques." *Ibid.*, vol. 16, nos. 1, 2, 1908.
408. ———. *Revue de Métaphysique et Morale*. Paris, 1900, vol. 9, pp. 169–188.
409. ———. "Novaya nauka i akademik Imshenetskiy" [The new science and Academician Imshenetskiy]. *Severnyy vestnik*, no. 12, 1896.
410. Erenfest, P. "Retsenziya na knigu Lui Kutyvra 'Algebra logiki'" [Review of L. Courturat's book 'The algebra of logic']. *Zhurnal russkogo fiziko-khimicheskogo obshchestva pri Peterburgskom universitete*, vol. 12, sec. 2 (1910).
411. Parfent'ev, N. N. "Retsenziya na knigu L. Kutyura 'Printsipy matematiki'" [Review of L. Courturat's book 'The principles of mathematics']. *Izvestiya fizichesko matematicheskogo obshchestva pri Kazanskom universitete*, vol. 18, no. 4, 1912.

412. Couturat, L. *Algebra logiki* [The algebra of logic]. (Russian translation). Odessa, 1909.
413. Sleshinskiy, I. V. *Dobavlenie k 'Algebre logiki' L. Cutyura.* [Supplement to L. Couturat's 'Algebra of logic']. Odessa, 1909.
414. Poretskiy, P. S. *Théorie des non-égalités logiques.* Kazan, 1904.
415. Curry, H. B. *Leçons de logique algébrique*, chap. 4. Paris, 1952.
416. Becker, O. *Einführung in die Logistik: Vorzügleich in den Modalkalkül*, Bonn, 1951.
417. Shestakov, V. I. "Ob odnom simvolicheskom ischislenii, primenimon k teorii releynykh elektricheskikh skhem" [On a symbolic calculus applied to the theory of electrical relay circuits]. *Uchenie zapiski MGU*, vol. 5, no. 23 (1944).
418. Sleshinskiy, I. V. "Pamyati Poretskogo" [Memories of Poretskiy]. *Vestnik opytnoy fiziki i elementarnoy matematiki*, no. 487, 1909.
419. Jorgensen, Jorgen. Reference 326, vol. 1, pp. 136, 142, vol. 2, pp. 42–100.
420. Styazhkin, N. I. "K voprosu o vklade P. S. Poretskogo v razvitie matematicheskoy logiki" [The question of the contribution of P. S. Poretskiy to the development of mathematical logic]. *Vestnik MGU—seriya ekonomiki, filosofii i prava*, 1956, no. 1, pp. 103–109.
421. ———. "Uproshchenie P. S. Poretskim nekotorykh algoritmov klassicheskogo ischisleniya vyskazyvaniy" [P. S. Poretskiy's simplification of certain algorithms of the classical calculus of propositions]. In *Logicheskie issledovaniya* [Studies in logic]. Moscow, 1959.
422. ———. Reference 329, pp. 3, 5, 10–12.
423. ———, and V. D. Silakov. *Kratkiy ocherk istorii obshchey i matematicheskoy logiki v Rossii* [Short outline of the history of general and mathematical logic in Russia], pp. 3, 4, 39–73. Moscow, 1962.
424. Bunitskiy, E. "Nekotorye prilozheniya matematicheskoy logiki k arifmetike" [Some applications of mathematical logic to arithmetic]. *Vestnik opytnoy fiziki i elementarnoy matematiki*, no. 258, 1896.
425. ———. "Nekotorye prilozheniya matematicheskoy logiki k teorii OND i NOK" [Some applications of mathematical logic to the theory of the greatest common divisor and least common multiple]. *Ibid.*, no. 274, 1899.
426. Volkov, M. S. *Logicheskoe ischislenie* [Logical calculus]. St. Petersburg, 1888.

C. S. Peirce (1839–1887)

427. Buchler, Justus, "Peirce's Theory of Logic." *The Journal of Philosophy*, vol. 36 (1939).

428. Wiener, Philip P. "The Peirce-Langley Correspondence and the Peirce's Manuscript on Hume and the Laws of Nature." *Proceedings of the American Philosophical Society*, vol. 91 (1947).
429. *Dictionary of American Biography* (New York, 1934), s.v. "Charles Sanders Peirce," by Paul Weiss.
430. Russell, F. C. "Hints for the Elucidation of Mr. Peirce's Logical Work." *The Monist*, vol. 18, 1908, pp. 406–415.
431. Ueyama, S. "Development of Peirce's Theory of Logic." *Science of thought*, vol. 1, 1954.
432. Young, F. H. "Charles Sanders Peirce: America's Greatest Logician and Most Original Philosopher" (Vogetr. am 15 October 1945 der Pike County Hist. Soc. in Milford, Penn.). Privately printed, 1946.
433. Feibleman, James. *Introduction to the Philosophy of Charles Peirce, Presented as a System.* London and New York, 1946.
434. Venn. J. *Symbolic Logic*, pp. 97, 134, 146, 189, 341, 383–384, 404, 407, 415. London, 1881.
435. Burks, A. W. "Peirce's Conception of Logic as a Normative Science." *The Philosophical Review*, vol. 52, 1943, pp. 187–193.
436. Reyes y Prosper, V. "Charles Santiago Peirce y Oscar Howard Mitchell." *El progreso matemático*, vol. 2, 1892.
437. Keyser, C. J. "A Glance at Some of the Ideas of Charles Sanders Peirce." *Scripta mathematica*, vol. 3 (1935).
438. Tannery, Paul. "Review of Peirce's 'On the Algebra of Logic.'" *Revue philosophique de la France et de l'Étranger*, vol. 12 (1881).
439. Townsend, H. G. "Review of 'Collected papers of C. S. Peirce, vols. 2, 3, 4.'" *The Philosophical Review*, vol. 43 (1934), pp. 209–212; vol. 44 (1935).
440. Eisele-Halpern, C. "The Theory of Probability in Charles S. Peirce's Logic and History of Science." *Abstr. Short communes International Congress Mathematics in Edinburgh*, 1958.
441. Scholz, H. "Review of 'Collected Papers of C. S. Peirce; vols. 3–5.'" *Deutsche Literaturzeitung*, vol. 57 (1936).
442. Peirce, C. S. *Memoires of the American Academy of Science*, vol. 9 (1870).
443. ———. "On the Algebra of Logic." *American Journal of Mathematics*, vol. 3 (1880).
444. Birkhoff, G. *Teoriya struktur* [Theory of structures], pp. 16, 36, 191, 389. Moscow, 1952.
445. Wells, G. *Pragmatizm—filosofiya imperializma* [Pragmatism: the philosophy of imperialism]. Moscow, 1955.
446. Moore, Edward C. *American Pragmatism: Peirce, James and Dewey.* New York, 1961.
447. Coudge, Thomas A. *The Thought of C. S. Peirce.* Toronto, 1950.

448. *Collected Papers of Charles Sanders Peirce.* Cambridge, Mass., vol. 5: *Pragmatism and Pragmaticism,* 1934.
449. Peirce, C. S. "Chance, Love and Logic." *Philosophical Essays.* New York, 1923.
450. *Collected Papers of Charles Sanders Peirce.* Cambridge, Mass., vol. 2: Elements of Logic, 1932.
451. ——— Vol. 3: Exact logic (Publ. papers), 1932.
452. ——— Vol. 4: The simplest mathematics, 1933.
453. Peirce, C. S. "On the Algebra of Logic: Contribution to the philosophy of Notation." *American Journal of Mathematics,* vol. 7 (1885).

J. Lachelier (1832–1918)

454. Lachelier J. *Du fondement de l'induction.* 1871.
455. ———. *Les conséquences immédiates et le syllogisme.* 1876.
456. ———. *La proposition et le syllogisme.* 1906.
457. Voyshvillo, E. K. "Kritika logiki otnosheniy kak relyativistskogo napravleniya v logike" [Critique of the logic of relations as a relativistic direction in logic]. *Filosofskie zapiski,* vol. 6, 1953.
458. ———. "Ob odnoy logicheskoy kontseptsii" [On one logical conception]. *Voprosy filosofii,* no. 6, 1957.
459. Povarnin, S. I. *Logika otnosheniy, ego sushchnost' i znachenie* [The logic of relations, its essence and significance]. Petrograd, 1917.

G. Frege (1848–1925)

460. Frege, Gottlob. *Begriffsschrift. Eine der arithmetischen nachgebildete Formelsprache des reinen Denkens.* Halle, 1879.
461. ———. *Über Funktion und Begriff. Berichte der Jenaischen Gesellschaft für Medizinische und Naturwissenschaftliche Forschung.* Jena, 1891.
462. ———. "Über Begriff und Gegenstand." *Vierteljahrschrift für wissenschaftliche Philosophie,* vol. 16 (1892), pp. 192–205.
463. ———. "Über Sinn und Bedeutung." *Zeitschrift für Philosophie und philosophische Kritik,* edited by R. Falkenburg. vol. 100 (1892), pp. 25–50.
464. ———. *Die Grundlagen der Arithmetik, eine logisch-mathematische Untersuchung über den Begriff der Zahl.* Breslau, 1884.
465. ———. *Grundgesetze der Arithmetik, begriffsschriftlich abgeleitet,* vol. 1, Jena, 1893; vol. 2, Jena, 1903.
466. ———. *Translations from the Philosophical Writings of Gottlob Frege.* Oxford, 1952.

467. Frege, Gottlob. "Über die Begriffsschrift des Herrn Peano und meine eigene." *Berichte der mathematischen Classe der königlich Sächsischen Akademie der Wissenschaften zu Leipzig*, 1896.
468. ———. "Was ist eine Funktion?" *Festschrift Bolzmann gewidmet zum 60-ten Geburtstage 20 Februar 1904*. Leipzig, 1904.
469. ———. "Kritische Beleuchtung einiger Punkte in E. Schröders Vorlesungen über die Algebra der Logik." *Archiv für systematische Philosophie*, vol. 1 (1895).
470. ———. "Über die Grundlagen der Geometrie." *Jahresbericht des Deutschen Mathematik Verein*, XV (1906). (A review of David Hilbert's book *Basis of Geometry*, published in Leipzig in 1899.)
471. Biryukov, B. V. "O rabotakh Frege po filosofskim voprosam matematiki" [The works of Frege concerning philosophical questions of mathematics]. *Filosofskie voprosy estestvoznaniya*, 2nd ed., *Nekotorye metodologicheskie voprosy fiziki, matematiki i khimii*. Moscow, 1959.
472. ———. "Teoriya smysla Gottloba Frege" [Gottlob Frege's theory of meaning]. In: *Primenenie logiki v nauke i tekhnike* [Application of logic in science and technology]. Moscow, 1960.
473. Yanovskaya, S. A. "Bstupitel'naya stat'ya k perevodu knigi Gil'berta D. i Akkermana V. 'Osnovy teoreticheskoy logiki'" [Introductory article to the translation of D. Hilbert and W. Ackermann's 'Fundamentals of theoretical logic'"]. *Osnovy teoreticheskoy logiki*. Moscow, 1947.
474. Scholz, H., and F. Bachmann, "Der wissenschaftliche Nachlass von Gottlob Frege." *Actes du Congrès International de Philosophie Scientifique*, VIII. Also in *The history of logic and the philosophy of science* (German), pp. 24–36. Paris, 1936.
475. Perel'man, Sh. "Frege's Metaphysics." *Kwartalnik filozoficzny*, vol. 14, 1938, pp. 119–142 (Polish, with French résumé). (Polish title: *Metafizyka Fregego*).
476. Beumer, M. G. "En historische bijzonderheid uit het leven van Gottlob Frege (1848–1925)." *Simon Stevin*, vol. 25, 1946–1947.
477. Wittgenstein, L. *Tractatus logico-philosophicus* (Russian translation: Moscow, 1958.)
478. Beth, E. W. Supplement to Reference 476. pp. 150–151.
479. *Filosofskiy slovar'* [Philosophical dictionary] (Moscow, 1963), s.v. "Frege."
480. Korcik, A. "Gottlob Frege jako tworca pierwszego systemu aksjomatycznego wspolczesnej logiki zdan." *Roczniki filosoficzne*, vol. 1, 1948, pp. 138–164 (Polish with résumé in French, p. 322).
481. Wienpahl, Paul D. "Frege's Sinn und Bedeutung." *Mind*, vol. 59 (1950), pp. 483–494.
482. Papst, W. *Gottlob Frege als Philosoph*. Dissertation, Berlin, 1932.

483. Black, Max. "A Translation of Frege's 'Über Sinn und Bedeutung,'" Introductory note in *The Philosophical Review*, vol. 57, 1948.
484. Linke, Paul F. "Gottlob Frege als Philosoph," *Zeitschrift für philosophische Forschung*, vol. 1, 1946–1947, pp. 75–99.
485. Bachmann, Friedrich. *Untersuchungen zur Grundlegung der Arithmetik mit besonderer Beziehung auf Dedekind, Frege und Russell.* Dissertation, Münster i W., 1934. Lithographed. See also *Forschungen zur Logistik und zur Grundlegung der exakten Wissenschaften*, no. 1, 1934.
486. Narskiy, I. S. *Sovremennyy pozitivizm* [Contemporary positivism], pp. 70, 148, 159, 179, 223, 336–338. Moscow, 1961.
487. Severi, F. (Review of a new edition of texts by Frege published under the title "Mathematics and logic"). *Scientia*, vol. 84 (1949), p. 144.
488. *Archimede*, vol. 1, no. 1 (1949), pp. 34–35. (Review of *Mathematics and Logic*).
489. Lukasiewicz, J. "Z historii logiki zdań." *Przegląd Filozoficzny*, vol. 37, no. IV, Warsaw, 1934.
490. Kneale, W., and M. Kneale. *The Development of Logic*. Oxford, 1962. See pp. 403, 413, 426–427; 435–512, 518, 519, 524, 528, 530, 537, 539, 587, 594, 602, 604, 622, 686, 738.
491. Church, Alonzo. *Introduction to mathematical logic* (Russian translation: Moscow, 1960, pp. 17–20, 341–372, 380–381, 401–402, 440, 450).
492. Biryukov, B. V. "Avtonomnoe upotreblenie vyrazheniy" [Autonomous use of expressions]. In *Filosofskaya entsiklopediya*, vol. 1.
493. ———. "Vzaimozamenimosti otnoshenie" [Interchangeability of relations]. *Ibid.*

V. Minto (1845–1893)

494. Minto, V. *Deduktivnaya i induktivnaya logika* [Deductive and inductive logic]. Moscow, 1905.

Precursors of Peano

495. Saccheri, G. *Logica demonstrativa*. Turin, 1697, Cologne, 1735.
496. ———. *Euclides ab omni naevo vindicatus*. Milan, 1733; London and Chicago, 1920.
497. Richeri, Ludovico. "Algebrae philosophicae in usum artis inveniendi specimen primum." *Mélanges de philosophie et de mathématique de la Société Royale de Turin*, vol. 2, chap. 3, 1761.
498. Pagano, F. *Logica dei probabili*. Napoli, 1806.

499. Galuppi, P. *Logica pura*. 1820.
500. ———. *Lezioni di logica e di metafisica*. 2 vols. 1832–1833.
501. ———. *Historisch-biographisches Handwörterbuch zur Geschichte der Philosophie*: Leipzig, 1879.

G. Peano (1858–1932)

502. Peano, G. "Notations de logique mathématique." Introduction to Formulaire de Mathématiques. Turin, 1894.
503. ———. *Revue de Métaphysique et de Morale*, vol. 8, 1900, pp. 589–591.
504. ———. *Calcolo geometrico secondo l'Ausdehnungslehre di Grassmann, preceduto dalle operazioni della logica deduttiva*. Turin, 1888.
505. ———. *Arithmetices Principia, Nova Methodo exposita*. Turin, 1889.
506. ———. *I principii di Geometria, logicamente expositi*. Turin, 1889.
507. ———. *Formulaire de Mathématiques*, 5 vols. Turin, 1895–1905.
508. ———. *Rivista di Matematica*, vol. 1, 1891, pp. 93, 94.

Followers of Peano

509. Burali-Forti, C. Sopra un teorema del Sig. G. Cantor. *Accademia Torino*, XXXII, 1896–1897.
510. ———. *Logica matematica*. Milan, 1894.
511. *Revue de Méthaphysique et de Morale*, vol. 8 (1900), pp. 591–594.
512. Vacca, Giovanni. "La logica di Leibniz." *Rivista di matematica*, vol. 8, 1902–1906.
513. ———. "Sui precursori della logica matematica. II. J. D. Gergonne." *Rivista di matematica*, vol. 6, 1896–1899.
514. Vailati, Giovanni. *Rivista di matematica*, vol. 2, 1892, p. 73.
515. ———. *Rivista di matematica*, 1900–1901, pp. 148–159.
516. ———. "Un teorema di logica matematica." *Rivista di matematica*, vol. 1, 1891, p. 103.
517. Padoa, A. "Ce que la logique doit à Peano." *Actes du Congrès International de Philosophie Scientifique*, vol. 8, 1936, pp. 31–37.
518. Padao, A. *Periodico di matematica*, ser. 4, vol. 18, 1938, pp. 228–236.
519. Caruccio, Ettore. "Spunti di storia delle matematiche e della logica nell'opera di G. Peano. 'In memoria di Giuseppe Peano.'" *Liceo scientifico statale*, 1955.
520. Barone, F. "Un'apertura filosofica della logica simbolica peaniana." *Ibid.*, pp. 41–50.
521. Geymonat, L. "I fondamenti dell'aritmetica secondo Peano e le obiezioni 'filosofiche' di B. Russell." *Ibid.*, pp. 51–63.
522. Pedro Pi Calleja. "La objection di Grandjot a la teoria de Peano del numero natural." *Mathematicae notae*, vol. 9, 1949.

523. Feys, R. "Peano et Burali-Forti, précurseurs de la logique combinatoire." *Actes du XI-éme Congrès International de Philosophie*. Brussels.
524. Cassina, U. Sul "formulario matematico" di Peano. "In memoria di Giuseppe Peano." *Liceo scientifico statale*. 1955.
525. ———. *Periodico di mathematiche*, ser. 4, vol. 17 (1937), pp. 129–138.
526. Ascoli, G. "I motivi fondamentali dell'opera di Giuseppe Peano." *Liceo scientifico statale*, 1955.
527. Segre, B. Peano ed il Bourbakismo. *Ibid.*, 31–39.
528. Cassina, U. "Storia ed analisi del 'Formulario completo' di Peano." *Bolletino Unione matematica Italiana*, vol. 10, no. 4 (1955).
529. Geymonat, L. "Peano e le sorti della logica in Italia." *Bolletino Unione matematica Italiana*, vol. 14, no. 1 (1959). See also Abstract 439 by N. N. Veselovskiy in *Referativnyy Zhurnal Matematika*, no. 7 (1960), p. 2.
530. Genkin, L. O. *O matematicheskoy induktsii* [On mathematical induction]. Moscow, 1962.
531. Bourbaki, N. *Outline of the history of mathematics* (Russian translation, Moscow, 1963).
532. Church, Alonzo. (Reference 491), pp. 310, 349, 377, 385–388, 539–541.
533. Orey, Steven. "Formal development of Ordinal Numbers Theory." *Journal of Symbolic Logic*, vol. 20, no. 2 (1955).
534. Dedekind, R. *Was sind und was sollen die Zahlen?* Braunschweig, 1888.
535. Abstract 4192 in *Referativnyy Zhurnal Matematika*, 1955.

Index